How to Get the Most Out of The Source℠

Bantam Computer Books
Ask your bookseller for the books you have missed

THE APPLE //c BOOK
by Bill O'Brien

THE COMMODORE 64 SURVIVAL MANUAL
by Winn L. Rosch

EXECUTIVE GUIDE TO THE EPSON GENEVA
by Marshall P. Smith

EXPLORING ARTIFICIAL INTELLIGENCE ON YOUR COMMODORE 64
by Tim Hartnell

EXPLORING THE UNIX ENVIRONMENT
by The Waite Group / Irene Pasternack

FRAMEWORK FROM THE GROUND UP
by The Waite Group / Cynthia Spoor and Robert Warren

HOW TO GET THE MOST OUT OF COMPUSERVE
by Charles Bowen and David Peyton

HOW TO GET THE MOST OUT OF THE SOURCE
by Charles Bowen and David Peyton

THE MACINTOSH
by Bill O'Brien

THE NEW *jr*: A GUIDE TO IBM'S PC*jr*
by Winn L. Rosch

ORCHESTRATING SYMPHONY
by The Waite Group / Dan Schafer with Mary Johnson

PC-DOS / MS-DOS
User's Guide to the Most Popular Operating System for Personal Computers
by Alan M. Boyd

PORTABLE CALC + PORTABLE SCHEDULER FOR
THE PORTABLE EPSON PX-8
by Howard Levine

PORTABLE WORDSTAR FOR THE PORTABLE EPSON PX-8
by Marshall P. Smith

POWER PAINTING: COMPUTER GRAPHICS ON THE MACINTOSH
by Verne Bauman and Ronald Kidd / illustrated by Gasper Vaccaro

SMARTER TELECOMMUNICATIONS
Hands-On Guide to On-Line Computer Services
by Charles Bowen and Stewart Schneider

SWING WITH JAZZ: LOTUS JAZZ ON THE MACINTOSH
by Datatech Publications / Michael McCarty

TEACH YOUR BABY TO USE A COMPUTER
Birth Through Preschool
by Victoria Williams, Ph.D. and Frederick Williams, Ph.D.

How to Get the Most Out of The Source℠

Charles Bowen and David Peyton

BANTAM BOOKS
TORONTO • NEW YORK • LONDON • SYDNEY • AUCKLAND

HOW TO GET THE MOST OUT OF THE SOURCE
A Bantam Book / October 1985

Throughout the book, the trade names and trademarks of many companies and products have been used, and no such uses are intended to convey endorsements of or other affiliations with the book.

The Source is a service mark of Source Telecomputing Corp., a subsidiary of The Reader's Digest Association Inc.

Book design by Nicola Mazzella.
Cover design by J. Caroff Associates, Inc.

For information address: Bantam Books, Inc.

ISBN 0-553-34228-2

Published simultaneously in the United States and Canada

Bantam Books are published by Bantam Books, Inc. Its trademark, consisting of the words "Bantam Books" and the portrayal of a rooster, is Registered in U.S. Patent and Trademark Office and in other countries. Marca Registrada. Bantam Books, Inc., 666 Fifth Avenue, New York, New York 10103.

PRINTED IN THE UNITED STATES OF AMERICA

HL 0 9 8 7 6 5 4 3 2 1

This is dedicated fondly to some noble teachers—Jack LeRoy, Nancy McClellan, Stella Sumpter, and Jesse Stuart—who gave us knowledge, inspiration, and the hunger.

Contents

Preface

It's been almost seven years now since we first started dreaming about The Source, the way immigrants-to-be think about the promise of a newfound land.

The opening salvoes of the microcomputer revolution were just beginning to be heard around the world when we had a fateful conversation with Bob Myers, a local businessman who also happened to be a member of the West Virginia Board of Education.

Now, Bob wasn't—isn't—a technical sort, but he always seems to have a good ear for tuning out hype and homing in on ideas with substance.

"Folks," he said, "we're going to have to figure out a way to get these new computers into the school systems. They're what this new age is all about." In the same breath, he said, "Do you know that over in Virginia some people have started something called The Source? It's a big computer system you can dial up on a telephone and get information—all kinds of information—from it with your home computer. What's more, you can use it to communicate with others around the country—maybe around the world."

Bob's comments sparked our imagination. But that's not surprising, considering our background. Growing up in the newspaper business, we naturally thought of The Source's promise of an instant person-to-person information network as a predictable extension of the high-tech revolution we were seeing firsthand in newsrooms.

But we found it was much more than that. In fact, The Source would become magic words to tens of thousands of computer owners because the system could be different and special to each one of them. The same computer system that brought instant news services into our homes could be used by someone else to keep up-to-the-minute on stock market developments, or to do shopping from home or to publish an on-line newsletter for a nationwide audience. And there never has been a dearth of new ideas and new uses. Source subscribers have always had ongoing discussions about better ways they've found to use the new medium.

It's appropriate, we think, that we should have first heard about The Source from someone like Bob Myers, a person who isn't connected to the computer industry or even particularly enamored of technology. He and so many others realized, even back then, that personal computers held the key to new kinds of communications. In recent years we've heard the question, "But what can you *do* with a computer?" We submit that since 1979 The Source has been an answer just waiting for that question to be asked.

We're proud to offer our services to help you learn about the exciting world of The Source, using a technique that's been tested in the field. In our previous book, *How to Get the Most Out of CompuServe* (Bantam Books), we introduced a learning aid called "on-line tours." We'll use that approach again to help you explore The Source. It works like this: instead of simply telling you how things look and act in this electronic community, we'll show you. Throughout the book, we'll invite you to log on to the system and see it for yourself with us at your side.

Readers of the CompuServe book told us this technique made it easier for them to familiarize themselves with the basics and to customize CompuServe for their needs quickly. Newcomers told us they could fire up their computers and modems and take their maiden voyage without feeling so alone. Experienced subscribers said the book improved their on-line skills. They could browse through the introductory tours and join us later as we headed in to new waters.

Before we begin, we'd like to take a moment to thank some people who helped bring us all together for these tours.

First, our wives—Pamela Bowen and Susan Peyton—who talked us through the writers' blocks, edited the manuscript and forgave us our transpositions.

Thanks to Nancy Beckman, Taylor Walsh and Leslie Spira at The Source for providing us with information and reading parts of the manuscript for accuracy. Thanks to Diane Worthington, PARTICIPATE's online services manager for The Source, Paul Bunnell and Dawn Johnson for providing us with detailed explanations on PARTI commands used on The Source.

Thanks, too, to Katinka Matson and Michele Martin at John Brockman Associates, Inc.; Susan Crouch who, believing that editing really means quality control, test drove the book, and to the team at Bantam: James Walsh, Jono Hardjowirogo, Frank O'Donnell and Kenzi Sugihara.

Finally, thanks to the many fine people we met on-line. Their enthusiasm about The Source kept our spirits high during the research and writing. They showed us their favorite electronic stomping grounds so we could share them with you.

Charles Bowen
Dave Peyton
June 1985

Foreword

John C. Dvorak

Hey, these guys are pretty good! That was my first thought last year when I saw the first Bowen and Peyton book entitled *How to Get the Most Out of CompuServe.*

Gee, I wish they'd do a book about The Source! That was my second thought. Then, gadfly actor and computer hobbyist Robin Williams was visiting my domicile, and he insisted that *How to Get the Most Out of CompuServe* should belong to him. He held it to his chest and said, "I need this book." At that point I realized that Bowen and Peyton had stumbled onto a hot property.

Their latest effort (this book you're reading) will be as much a classic as *CompuServe,* I'm sure. The ability to explain the complex with clear and concise wording is an art. Bowen and Peyton do it better than anyone.

Our information utilities are still in their infant stages. They're still a little expensive; they're still a little cumbersome. But, they hold and distribute enormous amounts of valuable knowledge.

Writers who push the capabilities of the information utility industry are doing us all a favor by demystifying the complex and encouraging its use.

Once people discover the niceties of conferences, electronic messaging, and public databases these systems will start to fulfill their promise—limitless information at your fingertips. As it now stands, only a fraction of microcomputer users subscribe to these systems.

This concept is still so new that none of the information utilities have discovered the key to successful marketing of their services. Do you market them like books, like cars, like software, like soap, like TVs, or what? Nobody knows.

Information utilities like to think that they are selling information. But they are really selling privilege. The privilege to roam through the databases just as one roams through the stacks of a large metropolitan library. The privilege is billed by the minute—it's time. This privilege can be very expensive if one has no idea as to what's going on. The user needs a road map, and books like this provide that map.

I've always suspected that the information utility business came close to the much sought-after marketing metaphor—the razor blade business. You know, a business where you give away the razor so people will buy disposable razor blades, or you give away a camera so people will buy lots of film. In the case of information utilities, you give away the information so people will buy lots and lots of "connect time." Some information purveyors have a hard time living with this idea. They are like the straight-edge razor manufacturer hoping all the way to the bankruptcy court that "people want a quality razor, not a junky disposable." So they concentrate on extra-cost quality databases that are so expensive you can't afford to use them. The money saved using a surcharged travel database, for example, when added to the cost of an airline ticket and hotel reservations might be a higher tab than a simple call to a travel agent. The best databases, for my money, are the ones created by the systems' users.

To further ignore the potential of these systems, the purveyors add surcharges to 1200 baud lines instead of encouraging their use. Few seem to realize that ten million users of a system at 1200 baud will generate more connect time than a meager 100,000 users straining under the limitations of 300 baud.

Even so, the systems are useful and made more so by books like this one. This book will save you money by saving you time. So relax and enjoy, knowing that you will be getting the most out of The Source.

John C. Dvorak
July 1985

How to Get the Most Out of The Source

CHAPTER 1

Getting Acquainted

It's the kind of deal we would have scoffed at when we were young journalism students coming up in the sixties.

You say you're going to bring the resources of five news services into my home so I can stay in touch with national and local events around the clock from my living room?

Yes. And you can use the same computer to get the latest stock quotes. And you can buy and sell stocks electronically.

Uh-huh.

Or you can write instant letters to friends in other cities, or even around the world in more than seventy-five different countries, and have them delivered in seconds. Or, you can form an on-line conference with friends or business associates and keep a complete written record of the proceedings as the meeting progresses.

Sure you can. And we suppose when this thing's not busy, it'll play cards with us on Friday nights?

Well, if that's what you want, yes. But we'd suggest one of the system's original computer games. They're far more challenging and . . .

Yes, we'd have been skeptical of such a pitch. After all, back in the sixties, most daily information was delivered to editors in newsrooms on Teletypes, one clacking letter at a time.

Unfortunately, some people think it still is.

Not everyone, of course. There are some who have found an edge, an inside track on instant information.

In business, they pick up on a trend a day or two before it's heard on the street.

Among their friends, they're the ones with the lowdown on new films and books and magazines. And at parties they're overheard saying, "If you want to hold off a day or so on that, I might know somebody who can give us some more information about . . ."

For many, that inside track is The Source, the sophisticated computer information network that has set the standards for a thriving new industry.

What you've heard about The Source may make it sound almost magical—that it can turn your computer into a super communications machine that's good for business and a lot of fun for home. Tens of thousands of computer users around the world hook up daily to The Source to "talk" to each other. They share a community pool of information, from news to gossip, from technically specific to philosophically general.

You also may have heard some not-so-magical things: The Source can be confusing and a little intimidating. And expensive. You're charged for each minute you're logged on, even when you're hopelessly befuddled.

Those are the things we heard when we set out to explore The Source a few years ago. We can tell you it's all true.

We were enthralled with the system. The Source is a class act. In some areas it was as if we stepped into a massive, but efficient, library where clerks sorted through thousands of records in seconds to hand us the one or two we were looking for. Then around the corner was a flea market where we could buy, sell, or swap everything from harmonicas to hard disks. And electronic department stores—we did Christmas shopping that year from home, without the yuletide tired feet and snow blindness.

We were also rather badly lost, though. Of course, we should keep that in perspective. We can lose our bearings in a parking lot, but at least there we can look for moss on the north side of a lamppost. The electronic landscape wasn't so clearly laid out.

All in all, the size of our canceled checks from that period testifies that (1) we were lost about 45 percent of the time and (2) we had a ball every minute.

With this book, we propose to help you sidestep some of the pitfalls we stumbled into. To do that, we'll give you something we didn't have—guided tours of the various parts of The Source.

That means this will be a different kind of tutorial book. Instead of sitting off at one side and telling you what you may find if you ever connect to this system, we want to be at your side when you log on. We'll discuss what you see as you see it on the computer screen. If there's one thing we've learned in our electronic wanderings, it's that the best way to learn about computer networks is to get your hands on the keyboard and start the engine.

Our goal is simple and pragmatic. To save you money by:

—Knowing where you're going before you get there, so that you can get the most out of each minute you're hooked up.

—Customizing The Source for your own needs. We don't know how much you know about other computer networks, but we think you'll find that The Source has remarkable and unique options that let you personalize the system.

We have high hopes for you. By the end of our visits together, you'll be well on your way to being an expert in using The Source. You'll have learned how to reconfigure this system so that it will appear The Source is operating just for you. It will recognize the kind of information and features you're interested in from the moment you log on because you will have filed that information away. You'll even be able to issue long commands by pressing a single key.

Do you realize, this is exactly how people who don't have computers think that all computers work? Won't it be nice for you to find a system that does?

By the way, who are you, anyway? Since you can't tell us about yourself, we're having to assume that you're a brand-new immigrant to these shores, that you've never before connected your computer to The Source. Our tours will start from that perspective.

But that doesn't mean there isn't room on the bus for those of you who already have some firsthand experience on The Source. We know some subscribers who have tried and failed to make this system sing. Even more common are the ones who have become comfortable with one or two areas and stay in that niche, not wanting to spend the time and money to investigate something around the block. If that's you, please, get aboard. You may find you need only skim the next two chapters or so, if you're already comfortable with how menus work and basically how to navigate the system.

On the other hand, even if you don't have a computer, you're not left out. True, our book will be more meaningful if it's by your side when you fire up a computer and log on. But on the tours, we'll be printing on the pages everything that would be displayed on the screen, so if you're more comfortable doing your touring from an armchair, that's fine, too.

SOME ASSUMPTIONS

While we're making assumptions about you . . .

We're going to have to assume you already know how to make your computer "talk," that you have a passing acquaintance with modems and communications programs and whatever add-ons your machine needs to communicate.

There are about a hundred models of personal computers being sold around the world. For each there are probably ten to hundreds of communications programs available. Obviously, we can't take up specific questions about software and hardware. However, your computer dealer should be able to answer many computer-specific questions about how you deal with this system.

For instance, the question we still most often hear about The Source and other computer networks is, "Can I get printouts of what I see on the screen?" The answer is yes—The Source is willing, if you have a smart communications program that knows how to do it.

The Source, incidentally, sells its own smart communications software. It's called *SourceLink* and it's available for many computers, including IBM PCs and the Apple II family. It's a program that deals with the system on a very intimate level. However, we're assuming that you're *not* using *SourceLink* on these tours. Because *SourceLink* interacts with The Source on such a heart-to-heart level, it makes the screen displays different from what they'd be with other software. If you have a choice between using *SourceLink* and another program, we'd urge the latter for these tours, just so we'll be seeing the system from the same perspective. (However, if you're using *SourceLink* and don't have a choice, don't worry about it. It's a fine program and a good choice on your part. Just realize that some of the displays in this book will appear a little different to you.)

This brings us to an important point about The Source. It deals with us all as equals—whether you're running a $10,000 mega-machine that's come down to you through the family or a $99 wonder you snatched up at a garage sale. It really doesn't care so long as you've agreed on the same terminal settings. The Source suggests you communicate at 8-bit words, no parity, 1 stop bit. (Don't panic if you don't know what that means. Just check the manual of your communications program and see the section dealing with "setting parameters" or perhaps "UART settings." Or check with your computer salesperson.)

Another assumption is that you'll be communicating at 300 baud. Many people are using 1200 baud these days. We do, too. But we firmly believe that it's wise to explore new territory at the slower baud rate, because it's cheaper. The Source adds a $3 hourly surcharge for 1200 baud communications. On our tours, we'll be stopping from time to time to explain something you've just seen or that you're about to see. No use paying extra money just to rev the engine.

WHAT IS THE SOURCE?

The Source, operated by Source Telecomputing Corporation, was introduced in June 1979 at the COMDEX computer trade show at a time when the market for personal computers was just getting started. The

Source is headquartered in McLean, Virginia, just outside Washington, D.C.

It was the dream of two men, William F. von Meister and Jack R. Taub. They originally considered an enterprise to deliver electronic mail by way of the unused portion of the FM radio signal, later deciding to use telephone connections instead.

In those days, The Source had no large computers of its own; it contracted for the space from ITT Dialcom of Silver Springs, Maryland, for use during off-hours. The system had only a small family of about two thousand users.

Still, the idea of a national network of computer users captured the attention of entrepreneurs intrigued by the romantic, still untested notion of "electronic publishing." In fact, those feelings were so strong they have earned The Source a permanent place in computer lore. Today, most computer users who have never even logged on to a computer network have heard of The Source. And if they haven't, they can read about it in frequent articles in newspapers like the *New York Times*, the *Washington Post*, and the *Wall Street Journal*.

A lot has happened since those early days.

For instance, in April 1980 *Reader's Digest* bought controlling interest in the fledgling network and has expanded the system. In 1981 the *Digest* authorized The Source to build a computer facility of its own and to purchase a bank of Prime minicomputers to take the place of those it used to lease.

The influence of *Reader's Digest* can still be seen, particularly as The Source continues to pursue that original goal of electronic publishing. For instance, The Source maintains an in-house, round-the-clock editorial staff that gleans news from various resources for on-line news bulletins and in-depth features.

In 1984 a big name in the computer industry, Control Data Corporation of Minneapolis, became a major investor in the network and contributed leadership as well as financial support.

The founders of The Source have moved on to other ventures and in 1984 a new president took over. He's Jay Keller, who formerly headed Control Data's industry and product strategy for the business information services division. Before moving to The Source, he was vice-president of electronic media development for Control Data's videotex division, Ticketron.

And all the while, The Source's on-line community has grown. By mid-1985, there were nearly 70,000 subscribers, with thousands of new arrivals checking in every month.

These days, The Source can be reached with a local phone call in about 1,400 U.S. cities and towns through packet networks like Telenet, Uninet and DataPac. There probably is a local telephone number in your area. In the next chapter we'll tell you how to reach The Source.

LET'S MAKE A DEAL (OR SEVERAL, ACTUALLY)

But first, a few words about the nature of this community you're entering.

Unlike the world of stone and glass and wood all around you, the on-line community of The Source is undergoing constant change. No sooner do you settle in a neighborhood on the outskirts of the system than a new feature springs up next door with slightly different commands and a different appearance.

In other words, if you think you can freeze The Source and put it under glass to study it, you're going to be disappointed. This industry still is very young. The system reflects that—new, better ideas for commands or features are always replacing the old ways.

What that means to us is this: even though we've tried to make this book as up-to-date as possible, there may be times when an example you see in the book isn't precisely what appears on the screen. In that case, someone at The Source has found a better way to do that job. In fact, The Source is proud that it has lines of communication open to you and all other subscribers for input on new ideas and for suggestions on how to improve existing features.

The way we should deal with this ever-changing landscape is to add another goal to our project. As important as saving time and money is, we think it is equally important to help you become self-reliant. We'd like you to be better able to roll with the occasional changes in the system's structure.

For our part, we'll try to help by carefully selecting the features we visit on our on-line tours. We're making no effort to stop at every single feature on the system. We'll choose areas that are representative, but more importantly, those that will teach you something about the nature of the system as a whole. (Of course, we'll also try to point out areas you can explore further on your own.)

For your part, we hope that if you come across a place where the system has changed, you'll look upon it as a challenge—not a frustration. In these rare cases, you can look at the example in this book to see how the feature used to work, then look at the screen to see how the programmers have improved it. Between the two, you'll be able to figure it out. And that's what becoming an expert in this territory is all about.

Finally, a thought or two about money.

In the interest of economy, please take your on-line tours only in "non-prime" time. That means evening hours during the week (after 6:00 P.M. local time), or anytime on Saturdays and Sundays. Rates are higher during business hours.

Second, as we've already suggested, use 300 baud. The rates are lower and it's smarter for covering unfamiliar ground. Save 1200 baud until you're familiar enough with the neighborhood and can offset the higher rates by knowing precisely where you're going.

Even with these money-saving tips, it's going to cost you a few bucks to take these tours. We figure we have about six hours' worth of on-line time scheduled for you. At the current rates (14 cents per non-prime time minute at 300 baud), this will add up to just about $50. Modestly, we suggest that's about 5 percent of what you'd probably pay to see all these areas of The Source if you used the bump-and-stumble method we used when we were starting out.

So here's the final deal: we promise to show just as much of The Source as we can in these six hours and, along the way, teach you some shortcuts that will save you time and money in the future, if you, in turn, promise not to be distracted by the clock.

CARRYING ON WITHOUT US

After we've finished the tours, we'll wrap up with a special section we think you'll find useful for future reference. The On-line Survival Kit in the back of the book is a collection of commands, features, lists, and tips. It should help you as you continue using the system long after we've been retired to the bookshelf.

And that's the bottom line, really. This book is a starting point and will show you firsthand enough about The Source to let you carry on the exploration on your own.

CHAPTER 2

How The Source Is Organized—Menus vs. Commands

If you took your old English professor's message to heart and spend a lot of time with your nose in a dictionary, you may already know this, but the word *menu* came down to us from a Latin word which means "diminished."

We usually think of menus in connection with restaurants, where that etymology makes sense, when you think about it. When you order from a menu, you "diminish"—reduce—the total bill of fare to the item you want. You don't need a menu if you want to tell the waiter, "I'll take one of everything." But you do need one to say, "I'll pass up everything but this."

According to the original Latin meaning, a highway map also is a menu of sorts. When you start out on a trip, you use a map to reduce the entire road system to a single highway or a series of connecting highways.

Of course, menus have taken on a new meaning in this age of high technology, and you've no doubt already met one on a computer screen. It's the most common way of directing you to the multiple functions of a program, to say, "Here are the things this program will do and here's the way to do them."

The Source has learned this ancient lesson, too. You can get to almost all the services The Source has to offer through menus. There's a "welcome menu," a "main menu," and eight submenus branching off the main menu. Also, most items on those branches have menus of their own. No wonder The Source calls its outline a "menu map," a valuable tool to get from one place to another in the system.

GETTING COZY WITH MENUS

The Source menus serve several important functions, especially for new users.

First, they remind you of where you are in the maze of features. For example, when this menu appears on your screen, you know you are at the "top" of the system, or just inside the front door:

****

MENU OF SERVICES <MENU>

1 Today From The Source <TODAY>
2 News, Weather and Sports <NEWS>
3 Business and Investing <BUSINESS>
4 Communications Services <COMM>
5 Personal Computing <PC>
6 Travel Services <TRAVEL>
7 Shopping, Games and Leisure <HOME>
8 Member Information <INFO> **FREE**

Enter item number, <H>elp or <Q>uit:

****

The last line here is called a *prompt*. You can choose the information or service you want by simply typing a number after the colon (:) and pressing your RETURN or ENTER key.

Menus also eliminate the need to remember exactly where everything is on The Source. When you're in a taxicab, you don't need to know your friend's specific address to find his house, so long as you can tell the driver, "It's three doors down from the corner." It's the same when you're navigating The Source with menus. Using this main menu, if you type 2 and press RETURN, you'll be taken to the area of The Source where you can retrieve news, wherever those services happen to reside in the massive computer system. Entering 6 <RETURN> takes you to the travel features branch, 4 <RETURN> goes to communication services, etc.

The downside of using menus is that it usually takes a while to get where you're going.

For example, choosing Option 7 on the main menu doesn't mean that the very next page will contain exactly what you're looking for (in this case, shopping features and games). In fact, more than likely it will give you another menu from which to choose a number, then perhaps even another menu after that. Each selection moves you a little closer to your destination, one level of menus at a time, sort of like moving from one branch to the next on a family tree.

When we start getting on-line in the next chapter, we'll show you where you can go on the system to pick up and print out a free copy of The Source's latest menu map, which shows all the major system menus. And, if you don't have a printer, stay tuned—you'll later learn how to send electronic mail to The Source's customer-service department and order a copy to be mailed to you.

Incidentally, menus *do* change from time to time. The Source initiated a whole new menu structure in early 1985 (in time for this book). And subtler changes in menus are going on all the time. When The Source adds a new feature, for instance, it often puts it as the first option on a menu, moving the older features down. For you, that means two things: first, don't be surprised if the menus you see in your on-line tours are occasionally slightly different from the ones we display here in our examples. And, second, it's a good idea to pick up a new copy of the menu map from time to time.

A BETTER WAY

By now, you may be getting a familiar sinking feeling. Since time is money on any dial-up information service, you're not sure if you want to spend your time watching menus scroll by while you imagine the meter running out of control.

Relax.

The menus are one way of getting around the system, but certainly not the best. There's an express that'll get you there quicker than you can go get a cup of coffee. Like a direct flight in an airplane, the express flies by all the menus and takes you directly to the service you want.

It's called navigating at the Command Level prompt and those words you see on the menus in brackets, <>, are the key. Our taxi driver will eventually find "three doors down from the corner," but he'll find it a lot faster if you know the house number. On The Source, those words in brackets are the house numbers.

In the next chapter you'll see how to get to the Command Level and how to use these addresses to zip around the system like a pro. Later, we'll show you how to link commands together to go even faster and make it appear as if the system is reacting automatically to your every wish.

Why, before it's all over, your friends'll be convinced you're a magician and no one will ever kick sand in your face again.

GETTING READY TO GET ON-LINE

Before we go on-line, though, we need to check your gear and how you've got it set up. Logging on to a new information service will always go easier if you prepare ahead of time. Think of this as a simple checklist you can use each time before you start one of the tours.

Make sure:

• It's non-prime time (that is, between 6:00 P.M. and 7:00 A.M. on weekdays, all day on weekends and holidays).

• Your modem's ready for 300-baud communications (to take advantage of the lowest possible charges).

• The parameters in your communications program are set properly—full duplex, no parity (or null), 8-bit words, 1 stop bit. (Consult the manual that came with your software.)

In addition, if you have a printer and your terminal program can print incoming information, turn it on for these early sessions. It can be a great study aid after you're off-line. Later, when you begin using The Source as a true information utility, you'll learn that sometimes nothing takes the place of the printed word, whether it's a printout of a news report, an electronic letter, or a help file.

THE TOOLS THE SOURCE GAVE YOU

When you signed up for the service, The Source gave you several pieces of information that are vital to getting you on-line:

1. The telephone number and the name of the communications network you should call to link up with The Source. This is likely a local number to a service called Telenet, Uninet, or DataPac. (If you don't have a link to one of these networks in your community, this may not be a local number.)

2. Your host computer system number. Your account is assigned to one of several mainframe computers at The Source. Yours probably is a number between 10 and 20. The networks that serve The Source use this code to route your call to the correct host computer.

3. Your user ID, the six-character code that identifies you as the owner of the account with that code. It's usually three letters and three numbers. Charlie's, for instance, is BBZ824; Dave's is BBP112.

4. Your password, which is a word or phrase that enables you,

and you alone, to control access to your account. The password can be changed as often as you'd like and the one assigned to you should be changed the first time you log in. We'll be doing that on the first tour.

About your password—never tell it to anyone. That's like giving a stranger your savings account book. And don't keep your password in the same place you keep your user ID. Just remember that anyone with these two pieces of information can log on to The Source and charge the usage to your bill. If you ever have reason to believe someone has taken your account number *and* your password, report it immediately to The Source by calling the customer-service number, 800-336-3330.

DIALING IN

When you call the telephone number you've been given for access, you ought to hear a high-pitched tone. That's the network's modem saying "hello." To make the connection, put your modem in "originate" mode. (Depending on the kind of modem you have, this could involve flipping a switch to "originate," or perhaps the modem itself will take care of that. Check your manual. The operative word here is "originate.")

What happens now depends on which network you're hooked up to.

If It's Telenet . . .

Press the RETURN or ENTER key on your computer twice. The word "TERMINAL" ought to appear on your screen. Press the RETURN or ENTER key again. Then you'll see an "@" character. This is a prompt where Telenet is looking for you to enter a C (for "connect") followed by the system number The Source gave you.

If your system is 10, type C 30124.
If your system is 11, type C 30138.
If your system is 12, type C 30147.
If your system is 13, type C 30128.
If your system is 14, type C 30149.
If your system is 15, type C 30148.
If your system is 16, type C 301156.
If your system is 17, type C 301159.
If your system is 18, type C 301162.
If your system is 19, type C 301158.
If your system is 20, type C 301408.

Type it exactly as shown, with a space between the C and the first number. Then press RETURN or ENTER after entering the last number.

You'll know the connection was successful when you see the system number appear, followed by the word "CONNECTED" on the next line.

If It's Uninet . . .

After the connection is made, the first prompt you should see is ">L?". Press RETURN or ENTER, followed by a period, then another RETURN or ENTER—remember it as <RETURN>. <RETURN>.

Next, you'll see something like:

****

UNINET
PAD XXX
PORT XX
SERVICE:

****

At the colon prompt, you should type the letter S, followed by your system number. If your system is 10, type S10 and press RETURN or ENTER. If your system is 11, type S11 and press RETURN or ENTER, etc. (Note that on Uninet, there is *no space* between the S and the number.)

If your connection was successful, you'll see something like the following:

****

*U001
OOO CONNECTED
to 70300000
CONNECTED TO
The Source
PLEASE TYPE ID
THEN YOUR
NUMBER AND
PASSWORD

****

If It's Uninet in Canada . . .

Uninet in Canada is accessed through DataPac. After dialing the local DataPac number, type a single period and press RETURN or ENTER if your modem is 300 baud; type two periods and press RETURN or ENTER if your modem is 1200 baud.

After making the connection with DataPac, you'll see the words "DATA PAC" and some numbers representing the port addrcss.

At that point, you type "Set 2:1" and press RETURN or ENTER. The "Set 2:1" will not be displayed. To connect to the Uninet network, type the following:

****

P 1 3125 70300005

****

Press RETURN or ENTER and you should see the following if the connection was made:

****

DATAPAC CALL
CONNECTED TO
1 3125 70300005
(09C) (1, P, REMOTE CHARG-
ING) PACKET
SIZE:128
CONNECTED TO
THE SOURCE

****

BEFORE WE HIT THE ROAD

Just a few more last-minute details before we hook up to The Source. First, about your keyboard. . . .

You may notice we've been telling you to "press RETURN or ENTER." In the future, all we'll be telling you is to press RETURN. You're a sharp person. You'll know that if your keyboard doesn't have a RETURN key, then we're talking about your ENTER key or your carriage return key.

Sometimes on The Source, you can use the BREAK key to interrupt what you're doing. However, some communications programs disable the BREAK key. If the BREAK key doesn't work (or if your keyboard doesn't have one), you can tap the CONTROL and P keys together to do the same thing.

Some computers don't have a key labeled "CONTROL." If that's the case, you need to find out from your user's manual or from your computer dealer what combination of keys will function as "control" on your computer. (On some older computers, for instance, holding down the SHIFT key and the down-arrow key signals control.) The CONTROL key will be very important while we're navigating The Source.

Sometimes the system will send you a message that you can use an "ESCAPE" key to perform a certain function. If your keyboard doesn't have an ESCAPE key, check the manual or call the computer store. There may be a combination of keys that will generate the escape code (such as a SHIFT up-arrow on some systems). At any rate, The Source usually will accept alternatives for the ESCAPE key, and they'll be identified.

Speaking of commands, as you go through the book, you'll notice that we type most of our commands in capital letters. We do that just to make the commands stand out in the printed examples. If your computer is capable of both capital and lowercase letters, you can type either for almost any commands. The Source isn't picky. And when it comes to writing text, both capital and lowercase letters are acceptable.

Finally, the biggest complaint we hear about The Source is that sometimes the system is "sluggish," that there seems to be an unusually long hesitation between the time you enter a command and the time the system does anything. This often isn't The Source's sluggishness but that of the network by which you're connected to it. But, of course, knowing that doesn't make it any easier to bear. As more and more of us want to look into these information services, the Telenets and Uninets of the world are having to host an awful lot of new visitors. You can do more to avoid the rush hours if you've got a little flexibility in your hours to connect to the system. For instance, if you always log on at the same hour and you almost always find the system slow, try it an hour earlier next time, or an hour later.

And here's a tip about that. Remember that weekday non-prime time on The Source always begins at 6:00 P.M. *local time*. That means that by 9:00 P.M. Eastern Time, every time zone in the continental United States has reached a non-prime-time hour. We've found most sluggishness occurs on the system between 9:00 P.M. Eastern Time and about 11:00 P.M. when the Easterners begin to tucker out and go to bed. If your schedule permits enough flexibility to avoid the peak times like that, you'll probably be able to sidestep the famous network slows.

CUE THE TRUMPETS AND DRUMS

You've now learned everything you need to log on to The Source the first time. Except this: be prepared to see some things you may not find in this book. As we've said, The Source is constantly adding features and services. The menus and commands you see in this book are likely to be about the same, but there could be some additions.

This medium is still quite young. Changes are to be expected and even looked forward to in this new way of obtaining and trading information. Learn to cope with any differences you may find. If you'll take these changes in stride, we can guarantee you'll become proficient at using the service in no time at all.

CHAPTER 3

Getting Our First Byte of The Source

The time has come for our first on-line tour, a chance to test your wings. If you've never seen The Source before, the best way to meet it is firsthand. You can read a *National Geographic* story about a country, but even the best reportage is pale compared with the real thing.

This tour will introduce you to some of the important services of The Source and give you a bird's-eye view of the system. It will also show you the major commands for navigating—getting from here to there in this massive computer system—introducing you to principles that you will build on in the rest of the book.

Get comfortable. Make a place for the book next to your computer, even if it means tossing an angry cat to the floor. You'll be reading text in the book and on the screen and tapping keys on the keyboard, sort of like following "Jane Fonda's Workout" tape without all the perspiration.

One more reminder: please don't worry too much about the money meter ticking away. It's true the tours are going to cost a little bit in connect time. But if you start out letting the expense be foremost in your mind, you'll not get the full benefit of the tours.

Enough preaching.

KNOCKING AT THE DOOR

In the last chapter we took you up to this point on the system:

****

Connected to THE SOURCE
>

****

So, review the log-on instructions in chapter 2 if you need to and get to this point before you read on. We'll wait.

By the way, we'll have to admit that ">" is a little disconcerting in its abruptness, but it's the prompt at which you enter the letters ID, followed by a space, followed by your Source user ID, followed by pressing RETURN. Like this:

****

> ID BBP112 <RETURN>

****

After that, you should get a prompt asking for your password. When you type it, you'll not see the word on the screen. This prevents anyone looking over your shoulder from learning your password, a protection for your account.

After typing the password, exactly as it was given to you, press RETURN.

(If you made a typing error in the password or the user ID, the system will find out the two don't match. Then you'll be summarily logged off The Source and taken back to the network prompt where you'll have to start over again. Don't worry. It happens to everyone from time to time. Just go through the log-on routine again.)

When you've entered the proper ID and password, you'll see something like this:

****

BBP112 (user 15) logged on Sunday, 03 Mar 85 16:48:40.
Welcome, you are connected to THE SOURCE.
Last log-on Sunday, 03 Mar 85 00:08:32.
(C) COPYRIGHT SOURCE TELECOMPUTING CORPORATION 1985.

New Features Online Now! Type NEW for More.
->

WELCOME TO THE SOURCE

1 Tutorial and Introduction <INTRO> **FREE**
2 Menu of Services <MENU>
3 Member Information <INFO> **FREE**
4 Today From The Source <TODAY>
(Q or QUIT will take you to Command Level.)

Enter item number, <H>elp or <Q>uit:

****

You may get more information than this. In the space after the copyright notice, The Source often places brief announcements about new services. You may be told that The Source or part of The Source will be unavailable at some time in the near future while the technical staff handles maintenance. If you are logging on at 1200 baud (not heeding our sage advice), you'll be told that you are using "an extra value service." That means it costs more than 300 baud.

The menu following the bulletins isn't the main menu we talked about in the last chapter. This is the "welcome menu." If you were entering a library, this would be the shelf of special books up front, before you get to the regular stacks.

Note that two of the selections here contain America's favorite word—**FREE**. We're going to one of these areas right away. Once there, the meter is turned off, which should help you relax and enjoy yourself.

By the way, there's an outside chance that at some point in the early part of this tour, you may see a line that says "CHAT FROM . . ." followed by another user ID. This is a message from another subscriber inviting you to talk with him or her. For the time being, though, we'll have to suggest you not try to respond. Just keep following the tour. Eventually the person on the other end will get the idea that you're new to these parts and don't know all of the language yet (which is true enough) and "hang up."

This may seem a little rude, but regulars of The Source's CHAT feature are used to it. About midchapter we'll show you a command that will automatically turn down CHAT requests to make sure you won't get such a request while you're learning the ropes. Then in the next chapter you'll see how CHAT works and how you *can* talk to other users. In other words, one adventure at a time.

CHECKING OUT THE FREE STUFF

Right now, we're going to the menu item marked "Tutorial and Introduction <INTRO> **FREE**" which is Option 1.

At the prompt on the bottom of the menu on your screen now, type the number 1 and press RETURN. You should then see a brief explanatory paragraph followed by a new menu:

****

Welcome to The Source, your information and communications network. In order to get the most from your sessions with The Source, take a few moments to read through the sections listed on the menu below. Enter a number and press RETURN. When you are finished with a section, you'll be returned to this menu. Remember, there is no charge for reading through these brief sections.

TUTORIAL AND INTRODUCTION <INTRO> **FREE**

1 A Description of the Services
2 The Source Tutorial <TUTOR>
3 How to Use Menus
4 Menus Map
5 How to Use Commands
6 How to Get Help
7 How to Control the Display
8 How to Sign Off From The Source
9 Select This Option to Go to
    The Menu of Services <MENU>

    Enter item number, <H>elp or <Q>uit:

****

Here's the first place where a menu change may have occurred since we wrote this book. As this book was going to press, The Source was planning a tutorial program for new users and those who need a "refresher course" in using The Source. It's likely that The Source tutorial will be located in this menu, perhaps at the top. If so, we urge you to make note of it and return to this tutorial later after you have read the book.

If you remember our discussion in the last chapter about interlocking menus, it'll be clear that this is a branch of the welcome menu you've just seen.

Now we're going to get a description of the major services on The Source by choosing Option 1 from this menu. We want you to read them, of course, but it's just as important that you note how text files are displayed on your screen. It works the same in the entire system as it does in this free area. And if you have trouble following the text on the screen, don't worry about it—we're duplicating the entire message in the examples here, where you can reread them now or later when you're off-line.

So type 1 and press RETURN and you should see something like this:

****

A DESCRIPTION OF THE SERVICES

The Services of The Source are grouped into eight categories, which you will see referenced on The Main Menu. Words in <BRACKETS> are the commands you can type in at Command Level, indicated by the arrow -> to display a menu of the services available within that category. Wait until you are positioned at Command Level to enter the commands.

TODAY FROM THE SOURCE <TODAY> Option 1

Includes the day's top news stories, business and sports news and feature articles, selected for your convenience by The Source's newsroom. You'll also find "What's New," an up-to-date list of new features on The Source.

NEWS, WEATHER & SPORTS <NEWS> Option 2

Here you will find the news publishers: United Press International, the Associated Press, The Washington Post, Scripps-Howard, United Media Features and Accu-Weather. You can search through these categories using key words or you can read selections of packaged articles.

-More-

****

When the screen displays the "-More-" sign, the scrolling should stop.

If you have a screen that displays 24 lines of up to 80 characters each, that will be exactly a screen's worth. If, however, you have a 24-line-by-40-column screen, you will have filled up two screens and half of the text material on this page has disappeared. In fact, if your screen display is anything less than 24 by 80, the top part of the text is no longer on your screen.

Don't worry about that now. We'll show you several ways to deal with this. Here's the problem at hand: The Source says there's more but doesn't say how to get it. There are two ways. One way is to type F (for Forward) and press RETURN. But the easier way is simply to press RETURN. You can press RETURN at any -More- prompt you come across on the system to go to the next page. Do that now and the text continues:

****

BUSINESS AND INVESTING <BUSINESS> Option 3

Here you will find the complete Investor Service of The Source, including real-time (or delayed) stock quotes, a powerful portfolio manager system, and the online execution and confirmation of stock trades. The service is complemented by a full line of reference and news sources, including historical stock analyses and an employment service.

COMMUNICATIONS SERVICES <COMM> Option 4

This is the doorway to begin your exchange with other members using SourceMail; more than 70 bulletin boards, through live "Chat," computer conferencing and use of The Source online Member Directory. You'll also find the offerings of Member Publishing here.

PERSONAL COMPUTING <PC> Option 5

A comprehensive compendium of personal computer-related reviews in abstract form is available for hardware, software and peripherals. You'll find reference to the personal computing bulletin boards and the member publications dealing with personal computing.

-More-

****

Another -More- prompt, requiring you to press RETURN again for page 3.

****

TRAVEL SERVICES <TRAVEL> Option 6

Find the latest airline schedules and fares; read hotel and restaurant reviews for establishments worldwide; peruse through bargain or exotic travel opportunities, and even book your reservations through an online travel club.

SHOPPING, GAMES AND LEISURE <HOME> Option 7

Connects you to online catalog shopping, dozens of games, movie reviews, horoscopes and more!

MEMBER INFORMATION <INFO> Option 8

Your **FREE** guide to the latest information of interest to members. Includes reminders for changing passwords and direct connections to STC's Customer Support and Billing departments. You'll find current rates for usage and storage here. And, it's all **FREE**.

-End-

****

That last message means, of course, that you're at the end of this particular text file. The next time you press RETURN, you'll go back to the INTRO menu you saw just before you embarked on this adventure. So, do that now.

ONCE MORE, WITH FEELING . . .

This exercise was more than just a sneaky way of getting The Source to do some of our work for us. While the contents of the file you've just read *will* give you a good overview of the system, we mainly wanted to show you that this is how most newcomers read files on The Source. The words scroll along, sometimes too fast to read; every so often, The Source comes to a halt to see if you want -More-.

What happens if it scrolls too fast for you to read or what if you want to reread a passage?

There are some good ways to exercise your influence over the way the material is displayed for you.

Let's go back to familiar territory for the next exercise. Type 1 and press RETURN again to see page 1 of the same text:

****

A DESCRIPTION OF THE SERVICES

The Services of The Source are grouped into eight categories, which you will see referenced on The Main Menu. Words in <BRACKETS> are the commands you can type in at Command Level, indicated by the arrow -> to display a menu of the services available within that category. . . .

****

And so on, until it stops with:

****

-More-

****

One way to pick up on something that scrolls by too fast is to have the system repeat a page of text. If F allows you to go forward, then it stands to reason that B should let you go "back" to the previous screen of text, right? It does.

So type B <RETURN> to see the last screen over again.

Well, that's one way of making sure you can read everything a screen has to offer, but probably not the best way. It would be better if you could stop the screen from scrolling while your eyes catch up with the text.

That's what we're going to do on the next page with CONTROL S and CONTROL Q. Remember we told you how important the CONTROL key is in The Source. Find it now. In a minute, when you press RETURN and start the words moving up your screen again, press CONTROL and the S keys together and the display should stop. To get it rolling again, press the CONTROL and Q keys together.

Okay? Fire when ready.

Press RETURN, then alternate the CONTROL S and CONTROL Q as you see the text that begins:

****

BUSINESS AND INVESTING <BUSINESS> Option 3

Here you will find the complete Investor Service of The Source, including real-time (or delayed) stock quotes, a powerful portfolio manager system, and the online execution and confirmation of stock trades. . . .

****

If all went well, you probably were able to stop and start the text three or four times before you reached the -More- prompt at the bottom of the page. Felt good, too, eh, stopping and starting The Source's computer dead in its tracks with a single command? Ah, a veritable sonata of power and poetry.

GETTING-OUT COMMANDS

No doubt you're tired of reading this same section—we are—so let's skip the remainder of it and go back to the menu. To do that, type M and press RETURN at the -More- prompt now on your screen. That will take you back to the menu you saw before.

This M command is an important one. In most features it will take you back to the main menu in that service. In this case, it's the one you left a few minutes ago:

****

TUTORIAL AND INTRODUCTION <INTRO> **FREE**

1 A Description of the Services
2 The Source Tutorial <TUTOR>
3 How to Use Menus
4 Menu Map
5 How to Use Commands
6 How to Get Help
7 How to Control the Display
8 How to Sign Off From The Source
9 Select This Option to Go to
    The Menu of Services <MENU>

Enter item number, <H>elp or <Q>uit:

****

This time we're going to read item No. 7, but we're going to read it nonstop, without the -More- inserted after every 24 lines.

Why would you want to do that? Because if you read really fast or you are sending the output from The Source to a printer for reading later, you really don't want the scrolling to stop.

To do that, type the following after the colon:

****

7 PRINT <RETURN>

****

You'll see a long section of text that doesn't halt until it reaches the -End- marker. Use your CONTROL S to freeze the text if it's coming in too fast; CONTROL Q to start it again.

Now, if you were reading that file (and not just playing with the CONTROL keys, as we were), you saw some information about another "getting-out" command—the BREAK key, which will let you get out of a file without having to wait for a -More- or an -End- prompt.

Let's give that a whirl. Try this: enter "7 PRINT" and let a few lines scroll onto your screen. Then press your BREAK key and see if that stops the file. (It should if your terminal program allows you to use your BREAK key.) If that doesn't work or if your computer has no BREAK key, type CONTROL P—that's just another way of telling The Source to "break."

You'll interrupt the text and get a message asking you to press RETURN to get back to the menu. Do that so we'll all be back to the menu.

Two important points about CONTROL P (or the BREAK key):

1. There's often a lag between the time you enter the CONTROL P and when the file actually stops. (The network you're communicating through is getting the BREAK message to The Source, and getting back a reply sometimes isn't instantaneous.) When this lag occurs, some new users get into a bad habit of tapping CONTROL P three or four times to "get its attention." Don't do that. You got its attention the first time. Entering the BREAK signal several times in a row is just going to delay you further. The computer acts on each one of the BREAKs it receives from you, trying to figure out what in the world you're doing.

2. The BREAK or CONTROL P doesn't always take you back to a menu. As we'll see later, it sometimes carries you to a special prompt or a

different function. But it always gets you out of what you're doing in a hurry.

CONTROL P is particularly important if you're lost and want to get back to something familiar. Think of it as an ejection seat to take you out of something you don't want to be in.

Everyone should be back to the following menu. (If you're not, enter the M command at the prompt on your screen.)

```
*************************************************************************

TUTORIAL AND INTRODUCTION <INTRO> **FREE**

1 A Description of the Services
2 The Source Tutorial <TUTOR>
3 How to Use Menus
4 Menu Map
5 How to Use Commands
6 How to Get Help
7 How to Control the Display
8 How to Sign Off From The Source
9 Select This Option to Go to
    The Menu of Services <MENU>

Enter item number, <H>elp or <Q>uit:

*************************************************************************
```

Make a mental note of this area. It's an excellent place to come—since it's free—if you want to practice some of the navigation commands we just used.

And, as you've probably noticed by now, this is where you can come to pick up your free on-line copy of the latest system menu map (see Option 4).

MEETING THE MAIN MENU

But not now. Instead, we're going to make a big jump to another part of the service, where we'll see the main menu you've heard so much about. It is this main menu from which almost all other menus spring and we'll be seeing a lot of it in the remainder of the tour.

Enter 9 at the prompt and you'll see:

****

MENU OF SERVICES <MENU>

1 Today From The Source <TODAY>
2 News, Weather and Sports <NEWS>
3 Business and Investing <BUSINESS>
4 Communication Services <COMM>
5 Personal Computing <PC>
6 Travel Services <TRAVEL>
7 Shopping, Games and Leisure <HOME>
8 Member Information <INFO> **FREE**

Enter item number, <H>elp or <Q>uit:

****

This should look familiar. You had a preview of it in the last chapter. Even though this is the most important menu in the whole system, it works just like the others you'll see.

Let's see what's here. Select Option 1, Today From The Source <TODAY>, and see something like this:

****

TODAY FROM THE SOURCE <TODAY>

1 Hour's Top News <BULLETIN>
2 Business Update <BIZDATE>
3 Sports News and Scores <SPORTS>
4 Today's Features <FOCUS>
5 What's New on The Source <NEW>

Enter item number, <H>elp or <Q>uit:

****

For those who like to keep up with the latest news, this is your area of The Source. The news you'll find in this area comes from The Source's own 24-hour newsroom and it's updated frequently throughout the day and night with national, international, business, and sports news as well as a special daily feature. The last item on this menu, "What's New on The Source," is where you'll find up-to-the-minute news about additions and changes to the system. We suggest you check "What's New on The Source" at least once a week.

Let's see. How about the top news of the hour? Enter 1 at the menu and you'll see a submenu like this (except the news headlines will be different, of course—unless you've figured out a way to turn the clock back to March 3, 1985).

```
*******************************************************************

BULLETINS From The Source

Sunday, March 3, 1985

 1 NEWS AT A GLANCE (Updated Throughout The Day)
 2 WEEK IN REVIEW
 3 Bush to Visit African Famine Areas
 4 Administration Official Predicts More Aid for "Contras"
 5 Zimbabwe Opposition Leader Turns Up in Capital
 6 Anti-Nuclear Groups Rally for Arrival of U.S. Ships
 7 Americans Avoided Taxes Through Swiss Accounts
 8 Half of New U.S. Jobs Will Be in 30 Metropolitan Areas
 9 Shultz-Ortega Meeting Seen as Political Plus
10 SOURCE POLL: Should Government Bail Out The Farmers?

Enter item number or Help

*******************************************************************
```

The first two selections give you synopses of news for the day and the week. Most of the other menu items will give you more in-depth stories concerning the day's news.

The last item is interesting, too. Often The Source surveys its members on issues of the day. As you can see, the hot topic in March 1985 was whether the government should bail out the financially troubled farmers. If you want to participate in one of the polls, just enter the menu number and you'll be prompted to answer a series of questions. Afterward you'll be returned to the Bulletin menu.

Let's take a break here while you read anything of interest. On any of the items you read, you should notice that the pages or screens of information are separated with the now familiar -More- prompt. Press RETURN to get the next page, and tapping RETURN at the -End- message will bring you back to the menu:

****

1 NEWS AT A GLANCE (Updated Throughout The Day)
2 WEEK IN REVIEW
3 Bush to Visit African Famine Areas. . . .
Etc.

****

You can also use the M command you've learned.
When you've read all the news you want, we'll carry on.

QUITTING . . . IS JUST THE BEGINNING

It's time for a new, important command. It's called QUIT and it's executed by entering a Q at nearly any prompt on The Source.

Don't be confused, however. The QUIT command doesn't mean you'll log off the service. Instead, it tells the system, "Quit what you're doing and zip over into the fast lane."

Try it. Enter Q at the prompt on your screen (it doesn't matter if it's at the bottom of the menu, a -More- prompt, or -End- prompt) and you'll see:

****

->

****

Uh, we didn't promise you bells and whistles, so it's not much to look at. However, this little arrow prompt signals you that you're at what The Source calls "Command Level." It's here that a lot of powerful things can be done quickly and efficiently. Think of it as being at the very heart of The Source system, the place where all the action takes place without any window dressing.

Since we're here, let's do a little housekeeping that we promised at the beginning of this tour. At the -> prompt, type these words:

****

-> CHAT -OFF <RETURN>

****

(Note that there's a space between the word CHAT and the hyphen.) The system should hesitate a second, then give you another -> prompt. It's ready for your next command.

You've just turned off the CHAT feature. From now on, if a subscriber invites you to CHAT, you'll not receive a "CHAT FROM . . ." message, but the other person will receive a message from the system saying you're unavailable at the moment.

Hope we've got you properly intrigued about CHATting. In Chapter 4 we'll take some time out to learn all about CHAT and how to use it. Stay tuned.

DRIVING FROM COMMAND LEVEL

Remember how nearly every menu selection you've seen has one word in capital letters and enclosed in brackets at the end of the line? The first menu we visited, where we got free information about the system, was called <INTRO>. The menu we just left was called <TODAY>.

These highlighted words are the command names for the menus or services. If you type that name at the Command Level, the system knows where you want to go.

For example, the main menu is known simply as MENU. So, at the Command Level prompt, type MENU (without the brackets) and press RETURN to see:

****

MENU OF SERVICES <MENU>

1 Today From The Source <TODAY>
2 News, Weather and Sports <NEWS>
3 Business and Investing <BUSINESS>
4 Communication Services <COMM>
5 Personal Computing <PC>
6 Travel Services <TRAVEL>
7 Shopping, Games and Leisure <HOME>
8 Member Information <INFO> **FREE**

Enter item number, <H>elp or <Q>uit:

****

Now type Q <RETURN> and you're back at the –> prompt again. Try it again. At the prompt, enter MENU.

How about that? You're on your first tour and already you've learned how to go into warp drive!

We'll be back here several times before we log off, but we still have a lot of exploring to do on this tour. We want you to see how far you can wander into the system with the menus; we want to write a short electronic letter, and we'll show you how to change your password. Most importantly, we'll pick up some new commands along the way.

In other words, let's get moving.

TO THE PITS!

Some features reside deep within the system. For instance, a game appropriately named the Pits is five levels removed from the main menu. Let's see how we would find it using the menus.

If you've entered MENU at the –> prompt, you should now be at the main menu:

```
*****************************************************************

MENU OF SERVICES <MENU>

1 Today From The Source <TODAY>
2 News, Weather and Sports <NEWS>
3 Business and Investing <BUSINESS>
4 Communication Services <COMM>
5 Personal Computing <PC>
6 Travel Services <TRAVEL>
7 Shopping, Games and Leisure <HOME>
8 Member Information <INFO> **FREE**

Enter item number, <H>elp or <Q>uit:

*****************************************************************
```

To start on this long road, choose Option 7 to see:

****

SHOPPING, GAMES AND LEISURE <HOME>

1 Shopping At Home <SHOP>
2 Game Arcade <GAMES>
3 2,000+ Movie Reviews <MOVIES>
4 Horoscopes, Soaps and More <BYLINES>
5 Games Bulletin Board

Enter item number, <H>elp or <Q>uit:

****

Let's browse a bit in the Game Arcade by typing 2 and pressing RETURN.

You should see a list of games, with the last menu item indicating there are "More Games" to be found on the next level.

Choose the last number on the menu to see more games. By now, you may be wondering if the games ever end on The Source. Well, yes, they do. In the next menu level. Type 9 <RETURN> and you'll reach the last branch of the longest menu road in The Source:

****

PITS AND MORE GAMES

1 Pits <PLAY PITS>
2 *Trek <PLAY *TREK>
3 Othello <PLAY OTHELLO>

Enter item number, <H>elp or <Q>uit:

****

Now that we're here, aren't we going to play a game? Not right now, pilgrim. You can do that later on your own time, if you'd like.

Meanwhile, we have another important command to learn. It's the P command which is reverse gear. It stands for "Previous," but what it really does is put the system in reverse and let you back out of a menu.

To see how it works, type P at the prompt and press RETURN. You should see the previous menu.

Well, we're out of the Pits and back into the Dungeon. Hey, this menu stuff is almost an adventure game in itself.

Type P <RETURN> again to rise yet another level.

****

GAME ARCADE <GAMES>

1 The Source Game Library <HELP GAMES>
2 Blackdragon <PLAY BLACKDRAGON>
3 Super Blackjack <PLAY SUPERBJ>
4 Castlequest <PLAY CASTLEQUEST>. . . .

Enter item number, <H>elp or <Q>uit:

****

Next, out of the game arcade and back into the shopping mall. Type P again and press RETURN.

****

SHOPPING, GAMES AND LEISURE <HOME>

1 Shopping At Home <SHOP>
2 Game Arcade <GAMES>
3 2,000+ Movie Reviews <MOVIES>
4 Horoscopes, Soaps and More <BYLINES>
5 Games Bulletin Board

Enter item number, <H>elp or <Q>uit:

****

And finally, type one more P and press RETURN to arrive at the Menu of Services again.

The P command will work virtually anywhere in the system, either at a menu or a program prompt. It will take you to the previous level of menus or prompts. Very handy.

But, whew! If you had to do that each time you wanted to play Pits, you might spend more money getting there than playing the game.

Fortunately, you don't. As you now know, entering the words in the brackets—in this case, PLAY PITS—at the -> prompt of Command Level puts you on the fast track.

Let's try that. From the Menu of Services on your screen, enter Q <RETURN> to get you to the Command Level prompt. Once there, type:

****

-> PLAY PITS <RETURN>

****

There's no doubt about it. The direct route is certainly more satisfying. You're zipping right to the front door of the game:

****

Welcome to "The PITS" Version 8.00
Copyright (c) 1980–1983 Jim Walters/Dave Broadhurst

Would you like an introduction?

****

Sorry, still no time to play, but now that you know how easy it is to get here, you can come back anytime after you read more about The Source's game library in a later chapter.

Right now, type CONTROL P after the question mark. That will take you back to the -> prompt.

VISITING THE ELECTRONIC POST OFFICE

The electronic mailbox here is called SourceMail. To get a feeling for where it resides in the system, let's go to the mail area through the menus. This also should show you how the menus can come in handy when you're looking for a service you've never visited before.

At the -> prompt, type MENU and press RETURN.

```
***************************************************************

MENU OF SERVICES <MENU>

1 Today From The Source <TODAY>
2 News, Weather and Sports <NEWS>
3 Business and Investing <BUSINESS>
4 Communication Services <COMM>
5 Personal Computing <PC>
6 Travel Services <TRAVEL>
7 Shopping, Games and Leisure <HOME>
8 Member Information <INFO> **FREE**

Enter item number, <H>elp or <Q>uit:

***************************************************************
```

Hmmm! Mail is communication, so type 4 <RETURN>.

```
***************************************************************

COMMUNICATION SERVICES <COMM>

1 SourceMail
2 Bulletin Boards
3 Chatting Online
4 Computer Conferencing <PARTI>
5 Member Directory <MEMBERS>
6 Members Currently Online <ONLINE>
7 E-COM and Mailgram Messages
8 Member Publishing <PUBLIC>
9 How to Upload Files to The Source

Enter item number, <H>elp or <Q>uit:

***************************************************************
```

Find "SourceMail" on the menu and select that option. (It's No. 1 in our example.)

****

SOURCEMAIL

1 Overview
2 Instructions
3 Mail Check <MAILCK>
4 Begin Mail <MAIL>

Enter item number, <H>elp or <Q>uit:

****

As you can see, from this menu you can get an overview of The Source's electronic mail service; you can get on-line instructions; you can take Option 3 which allows you to see how many letters are in your mailbox, or you can go directly to an action prompt where you can actually send mail and read letters you've received.

Choose 4 and press RETURN. Here's what you'll see:

****

<S>end, <R>ead, <SC>an, <D>isplay, or <Q>uit?

****

Don't worry about what all these commands mean right now. We'd like you to <S>end a letter to yourself, so that's the only one we need to be concerned with for the time being. Type S <RETURN> and you'll see a short prompt.

****

To:

****

The program doesn't want a name. It wants a user ID. Since you're sending this letter to yourself, type your own user ID here. Unlike the procedure for logging on, you don't need to type "ID" in front of it. After you type your ID, press RETURN. You'll see the next prompt.

****

Subject:

****

You can type any subject header you like here, as long as it's not more than 78 characters. Let's call it "My First SMail." So, type it at the subject prompt.

****

Subject: My First SMail <RETURN>

****

Next you'll see a prompt which says "Enter text," then the program will take you to the beginning of the next line where you can start typing your message. You can type up to 78 characters before pressing the RETURN key. (Be sure to press the RETURN key at least every 78 characters. Tapping <RETURN> before the end of each screen line is a good habit to get into.)

Let's all type the same letter the same way, like this:

****

Enter text:
This is my very first SourceMail electronic letter <RETURN>
and I have addressed it to myself.<RETURN>

****

When you finish, type a period followed by an S on a new line, like this:

****

.S <RETURN>

****

That lets the system know you're finished and it can send the mail to the address you specified at the To: prompt earlier.

If all went well, you'll receive a message that the SourceMail (SMail) was posted, something like this:

****

Wait . . .
Sent to – BBZ824
<S>end, <R>ead, <SC>an, <D>isplay, or <Q>uit?

****

Your letter will go through the electronic circuits and be delivered to your mailbox. It's a wonderful way to communicate instantly with friends across the country or even elsewhere in the world. Regardless of distance, SourceMail can get your words to them in minutes.

We'll leave your letter in the mailbox for the time being and use it in our next on-line tour in Chapter 5. There, we'll learn about how to read and store SourceMail.

Let's get back to the Command Level by entering Q at the prompt.

CHECKING THE CHARGES

Now we're going to another free area of The Source where you can find the latest scoop on what all these electronic ramblings cost.

You could reach the section on rates and charges by following the Member Information <INFO> branch of the main menu. But we'll bypass a layer of menus and go right to it.

Incidentally, this is timely, important information, so we'd suggest you print it out if you have a printer.

At the –> prompt, enter RATES to see:

****

RATES AND STORAGE FEES <RATES> **FREE**
Effective Since October 1, 1983

1 General Rate Information
2 Foreign Rates
3 Storage Rates

4 Rates for Value-Added Services
5 Rates for Mailgram and E-COM Messages
6 Payment Terms
7 Free Services

Enter item number, <H>elp or <Q>uit:

****

Choose each option, beginning with 1. After each 24 lines of 80 characters, you'll get the familiar -More- prompt. Press RETURN to see more. When you get to -End-, press RETURN to get back to the Rates menu. Use CONTROL S to stop the scrolling and CONTROL Q to start it again. If you're sending the output to a printer, you may want to use 1 PRINT, 2 PRINT, etc., for that continuous scroll we mentioned.

When you've read it all, type Q at the colon prompt to get back to the command prompt.

PASSWORD SECURITY

We're almost done. How long have we been on-line, anyway?

The Source knows. At the –> prompt on your screen, enter the word TIME and the system will display your connect time.

Need to know the date? Enter DATE at the prompt and you'll be shown the date, day, and the present time.

Cute. But we have one more important task to do before we call it an evening.

We've mentioned you should change the password you were issued. Let's do that now.

First, think of a new password. It should be at least 5 or 6 letters long. While punctuation can be used in the password, the password can't begin with a punctuation mark.

When you have the new password in mind, type the letters CPW at the prompt. (That stands for Change Password.) You'll then be asked to type in your current password. The next prompt says:

****

New Password?

****

Type the new password here. As you type it, it won't be displayed on the screen. When finished, press RETURN. Next you'll be asked for confirmation:

****

Re-enter new password for confirmation?

****

If you don't enter the new password exactly the way you entered it the first time, you'll be told the password change will fail and you'll be taken back to the command prompt. If it was successful, you'll simply see the command prompt again. The next time you log on, you'll need to use your new password at the password prompt.

By the way, you can also save a little time here by stringing some of the commands together. For instance, you could enter CPW followed by your old password and the system would then prompt you immediately for the new password. But more about command stringing in our next on-line tour.

LOGGING OFF

Had enough? You should now be at the command prompt which is where you have to be before you can log off the system.

Never simply disconnect your modem from the phone line or turn your machine off. If you disconnect without typing OFF and pressing RETURN at the command prompt, you can be charged for extra minutes on the service.

So type OFF and press RETURN now. Your exit will look something like this:

****

–> OFF <RETURN>
BBP112 (user 15) logged out Sunday, 03 Mar 85 17:20:00.
Time used: 00 h 59m connect.
Wait . . .
Bye.

****

If you want to keep track of your time on-line, now's the time to make note of the date and time used.

In most instances, you'll be taken back to the network prompt through which you entered. At that point, you can disconnect your modem from the phone line.

SUMMING UP

In this first tour you've seen the two faces of The Source—the Menu Level and the Command Level. As you saw, they are closely related; one is sort of the "flip side" of the other. When you're browsing or searching for some specific service, you'll probably prefer the menu road. When you know what you're looking for, you can take the direct approach at the -> prompt.

We also introduced you to some important navigation commands:

RETURN allows you to continue to the next page, the next prompt, or the next portion of a list. The RETURN key also allows you to execute a command at a prompt.

P, which stands for Previous, displays the previous menu or prompt in most instances.

M displays the main or top Menu of the service you are using.

F, which stands for Forward, displays the next page. By the way, if you follow it with a number, such as F3, it will display the third page ahead.

B, for Back, redisplays the page just shown.

Q, for Quit, when used at most prompts, takes you to the Command Level.

PRINT allows you to scroll text nonstop to the end without seeing the -More- prompts.

OFF typed at the Command Level logs you off the system.

BREAK or CONTROL P both interrupt what you're doing and generally will either take you back to a menu, main program prompt, or to the Command Level, depending on where you are in the system.

We also showed you the value of the two other CONTROL codes:

CONTROL S stops a scrolling display.
CONTROL Q starts it scrolling again.

CHAPTER 4

CHAT, the Real-Time Conference Utility

The Source has lots of ways for us to talk to each other. For some subscribers, none holds more fascination than CHAT, a program that lets you communicate in "real time" with anyone who happens to be on-line the same time you are.

First, we ought to define *real time,* a computer term that seems to mean different things to different people. Let's start by saying what it isn't.

It isn't real-time communications if Dave sends Charlie a message through SourceMail this evening and Charlie reads it tomorrow evening and sends a reply that Dave might read day after tomorrow. The messages will wait in our electronic mailboxes until we get on-line and read them, whether that's tonight or next week.

There are obvious advantages to this form of communications. Like any letter, the SourceMail can be carefully thought out and organized ahead of time. It can be as long as we want. And, as you'll see later, electronic mail can even be written off-line with a word processor and transmitted to The Source to save money on connect time.

The drawback, of course, is the delay. The SourceMail program can get a message from the writer to the recipient's mailbox in a matter of minutes, but that's a delay, nonetheless. That some messages can't tolerate any delay is the reason telephones and mailboxes have coexisted so happily for all these years.

CHAT is The Source's telephone. It lets you and someone else on-line "talk" together at the same time, just as if you were talking on the phone, except, of course, your words are typed on the screen.

USING CHAT

It couldn't be simpler.

It all takes place at the Command Level –> prompt. Let's say you know that a friend with user ID BBZ824 is on-line at the same time you are. To send your friend a signal that you'd like to CHAT, you would type this at the command prompt:

****

–> CHAT BBZ824 <RETURN>

****

After a brief hesitation, the system would show you this message:

****

Please remember that entering your password is required only at sign on, and never in CHAT.
Enter:

****

Are we *still* harping about this?

Yep. When it comes to password security, we're going to be like mother hens. And obviously, so are the managers of The Source.

And that's particularly important when it comes to CHAT. More than one would-be thief has used CHAT to try to trick unsuspecting users out of their passwords. Their ploy is to lure neophytes into a CHAT, then transmit simulated, official-looking "systems messages," such as "PLEASE ENTER YOUR PASSWORD."

Fortunately, you already know that once you've logged on to the system, you never again have to enter your password during that session (unless you've selected the Change Password option we talked about in the last chapter).

But a reminder never hurts, because the consequences of carelessness are so great: the person who has your user ID and your password has your Source account.

MAKING AND ACCEPTING INVITATIONS

Now, about that "Enter:" prompt.

Following the password warning, the system is giving you an opportunity to type a brief greeting to let your friend know you'd like to converse. It's usually something short and simple, such as, "Hi. This is Dave. Want to chat a minute?"

Always press RETURN after typing your message. This will send it to the other user.

Meanwhile, what's happening on the other end of the phone?

Well, no matter what BBZ824 is doing, he'll see an interruption on the screen (assuming, of course, he hasn't turned off the CHAT function—more about that in a minute).

The interruption will look something like this:

```
*****************************************************************

   ** CHAT FROM BBP112        ****

   Hi. This is Dave. Want to chat a minute?

*****************************************************************
```

After the message, the system will return BBZ824 to whatever he was doing before the interruption. If he is reading an article, the text will begin scrolling again. If he's playing a game, he'll pick up exactly where he left off. If he's at a prompt of some kind, the cursor will move down a few lines to make way for the CHAT request, but then he can continue just as if he were still at the prompt. In other words, once the CHAT invitation is delivered, the system continues as if it never existed.

If BBZ824 decides to accept your invitation, he must first go to the Command Level -> prompt (by entering Q for QUIT if he happens to be in the middle of a program or somewhere on the menu roads). Once there, he could make contact by entering:

```
*****************************************************************

   -> CHAT BBP112 <RETURN>

*****************************************************************
```

So, notice this: if he intends to answer the CHAT request, he needs to take note of the caller's user ID when he receives the initial invitation. He'll need that to pick up the phone.

After entering his CHAT command, he'll get the same warning about passwords that you did and the Enter: prompt. At this point, full contact is made and the conversation can begin.

But how do you find out whether your friend's on-line at the moment? And what happens if he's turned off CHAT?

Uh, patience. We'll be getting into those and related questions in a moment.

And speaking of patience, you'll need some when you're the one extending the CHAT invitation. As you've seen, there's no way for you to know what the invitee is up to when he receives your request. He may be toward the end of a program or article and available to talk in a minute or so. On the other hand, he may be just starting a feature that could tie him up for the next five or ten minutes. He may decide *not* to stop whatever he's doing, particularly if a CHAT with you wasn't scheduled ahead of time. That would mean you could be sitting in CHAT mode waiting for someone who isn't coming.

Hmmm—let's add something to our Things-To-Do list: we need to get some tips on how to avoid being left in the electronic lurch.

But first . . .

SAYING YOUR PIECE

Assuming you *have* made contact with your friend, typing messages to each other is quite simple. CHAT isn't a stickler for length—we've found we could send some pretty long lines to each other through the system. However, it's a good idea to get into the habit of typing about one line on your screen, tapping the RETURN key, typing another line, pressing RETURN again, etc.

Now, there can be problems in this form of communication. For instance, sometimes people talk at the same time, interrupting each other's messages on the screen. There needs to be a way to say, "There. I've finished this thought. Now you talk." The Source manual suggests this protocol: when you have finished a complete thought, press the RETURN key two or three times to create several blank lines. That will let the person on the other end know you're done and it's his or her turn.

There are other ways. For example, we like to end each line of an incomplete thought with three dots (. . .) just before pressing the RETURN key. Then, when we've finished with what we had to say, we type a period and the letters "ga," which generally is recognized to mean "go ahead."

Thus, to the other user, the message might look something like this:

****

I need to talk to you a second about the new . . . (pause)
word processing program you got for your IBM PC . . . (pause)
Does it have a function for search and replace? . . . (pause)
And if so, is it easy to use? ga

****

When the person on the other end sees the "ga," he knows you're waiting for a reply. If this, or some other, protocol can be established early in the conversation, it will simplify things a lot.

Finally, in CHAT, breaking up is easy to do. When you want the conversation to end, all you do is enter CONTROL P or tap your BREAK key. If either of the two talkers does that, contact is immediately broken and both subscribers are returned to the -> prompt of Command Level.

TAKING THE PHONE OFF THE HOOK

CHAT can be a lot of fun, not to mention a most valuable instant communications tool. But receiving an unexpected CHAT request also can be a real pain if you're doing something on-line that requires concentration. So there needs to be a way to post a "Do Not Disturb" sign. You already know—there is.

In the last chapter we showed you how to turn off the CHAT while we were in the first tour. At the Command Level prompt, you entered:

****

-> CHAT -OFF <RETURN>

****

We'll be doing that at the beginning of each on-line tour from now on so you can concentrate on learning the system without interruption. If someone tries to contact you through CHAT after you have entered that command, he or she will see:

****

NOT AVAILABLE FOR CHAT

****

You, on the other hand, won't even know he tried to call. Once CHAT is turned off, the system doesn't let any of the calls through. And CHAT will remain off for that session, or until you intentionally turn it back on by entering the command:

****

–> CHAT -ON <RETURN>

****

(Note that in both the ON and OFF commands, there's a space between the word CHAT and the hyphen.)

But wait a minute. How would the caller—or you, for that matter—know in the first place who's on-line at the moment?

The Source will give you that information if you ask. At the Command Level prompt, you can enter:

****

–> ONLINE <RETURN>

****

After a few seconds, you'll receive an alphabetized list of the user IDs of everyone who's logged on. No names are displayed, only user IDs.

Wouldn't that be an awfully long list? You bet. On a busy night it might take several minutes to get the complete rundown. That's why we advise using that command sparingly. Still, if you really want to know, the ONLINE command is an excellent way of determining if a friend is on the system. And, as with most features on The Source, you can enter CONTROL P or BREAK to stop the list after it begins.

ETIQUETTE, PROTOCOL, AND GOOD MANNERS

Okay, now that we've explained how to CHAT, we're going to urge you not to rush to get on-line and try it out on the first unsuspecting subscriber you can find by picking random user IDs from the ONLINE list.

Not everyone who comes to The Source is interested in electronic conversation. To many—whether they're on-line for research, business information, or a little serious game playing—unwanted CHAT requests are akin to phone calls from cemetery lot salesmen.

And they can strike fear in the hearts of newcomers.

Picture it—it's the user's first time on-line. He's not certain about what to expect anyway and the routine suddenly is broken with a CHAT request. What's he supposed to do? He doesn't know anybody in this neighborhood. Now the phone's ringing and he doesn't even know how to answer it. And if he had found out how to answer, what would he say? And how would he hang up? It's almost enough to cause an anxiety attack.

Even worse—when you're chatting with someone on the system, you still can receive a CHAT request from someone else. Talk about indecision. What do you do? Hang up on the first person? Most likely you'll just ignore the second unsolicited call.

Now, don't get us wrong. We're certainly not down on real-time CHATs. They're fun and interesting and sometimes rewarding. We've known subscribers who've used CHAT regularly to play chess, to have romantic encounters (electronically, of course), to even conduct quick business conferences and negotiations. As writers, we've used real-time conferences to interview people for books and articles.

But we think you'll find that the secret to most successful real-time communications is planning. Scheduling your CHATs in advance means you won't be sitting around waiting for your friends to come on-line. Also, if the person on the other end is expecting your call, you'll be less likely to be left in the electronic lurch when you send the CHAT request.

GETTING PRACTICE AND ALTERNATIVES TO CHAT

Obviously, you'll need to practice your CHAT skills, but please don't do it by ambushing someone from the ONLINE list.

If you have friends nearby who are also Source subscribers, how about contacting them? Perhaps one of them will come on-line to help you practice. As you've seen, it's not a very complicated feature (it's one of the easiest ones we'll see here, in fact) and within a half hour, you'll be a CHAT expert.

On the other hand, if you don't know any other Source subscribers in town, the system itself can help you find a partner. Because of the variety of ways you can communicate with Source users, establishing times to meet and CHAT is relatively easy.

You've already had a little taste of one of the best—SourceMail. After the next chapter you'll know more about this powerful electronic mail program.

In addition, we'll tell you how to find other Source members with similar interests and backgrounds through a Member Directory and how to contact them through SMail.

And if you'd like to take your chances on a spur-of-the-moment CHAT, we'll show you how to place a message in the POST bulletin board, alerting readers to the fact that you're on-line and you'd like to CHAT. After posting the message, it's a simple matter of waiting to see if someone contacts you.

And, as we suggested at the start of this chapter, that's only the beginning of the communications features available here. Later on you'll meet PARTICIPATE, a powerful conferencing system that allows you to talk with 2 or 200 or 2,000 other people. It's not real-time talk, but we think you'll see how PARTICIPATE may be even better for many applications.

After that, you'll be looking at the Member Publishing and a feature called Sharefiles where you can construct your own soapbox and become an information provider.

In other words, the talking has only just begun.

SUMMING UP

In this chapter we looked at the commands used in the CHAT feature, all of which are entered at the –> prompt of Command Level:

CHAT, followed by a user ID, invites another subscriber to talk. The same command is used to accept a CHAT request sent to you.
ONLINE lists the user IDs of all other people on-line at that moment.
CHAT –OFF turns off the CHAT mode so that you will receive no CHAT requests.
CHAT –ON turns on the CHAT mode again after it's been turned off.
CONTROL P or BREAK terminates a CHAT session and returns both talkers to Command Level.

CHAPTER 5

In Search Of . . .

The last time we were on-line together, you saw the difference between menu and command modes and you learned the major navigation commands that will take you virtually anywhere in the system. From here on out, we'll be building on that information.

In this chapter you'll need your walking shoes again—we'll be going on-line to take a closer look at SourceMail. If you had only a passing interest in The Source before we met, you may have at least heard that The Source's electronic mail program is first-rate. This should show you why. You'll see how to retrieve mail and how to stash it away in your electronic filing cabinet.

Speaking of filing, this will be our first look at an important area called the "User File Directory." There you can store text files and even programs. But it's much more useful than that. Before we finish the tours in this book, you'll be using the UFD area for some pretty classy tricks.

And you'll start learning those tricks on this trek, as we introduce you to "command strings" and how to put two or more commands all together on the same line to save lots of time.

We're saving the best for last. Using some of The Source's excellent search features, you'll command a major wire service to look for news stories of interest to you. You'll also search an on-line registry to find out about other Source subscribers and how to contact them. What you learn here about searching will be useful when you explore other features on your own.

GETTING ON-LINE

It's going to be a busy evening, and we'll start out with something new—a slightly different way to log on.

Please call The Source access number now, make your modem connection, and type in your system number, as you did in Chapter 3, but stop short of typing in your user ID and password.

Previously, we've suggested that you type in your user ID, press RETURN, and wait for the system to prompt you for "Password:". That's a good, secure way of logging on because it "masks" your password—it isn't displayed as you type it.

However, you can also put the user ID and password together on the same line. After the > prompt, you can type ID followed by the user ID and a space and then the password, as in:

```
****************************************************************

> ID BBZ824 PASSWORD <RETURN>

****************************************************************
```

If you'd like, try doing that this time (as long as you don't have a pirate peeking over your shoulder). The obvious disadvantage to stringing the user ID and password together is that this way the password *is* displayed. Since we often demonstrate The Source to acquaintances, we don't like to get into the habit of using this method. But, to each his own.

Either way, after you log on you'll see the welcoming message, the date and time, and then this familiar menu:

```
****************************************************************

WELCOME TO THE SOURCE

1 Using the Source <INTRO> **FREE**
2 Menu of Services <MENU>
3 Member Information <INFO> **FREE**
4 Today From The Source <TODAY>

(Q or QUIT will take you to Command Level.)

Enter item number, <H>elp or <Q>uit:

****************************************************************
```

Let's go directly to the Command Level this time by entering Q at the prompt. There we'll turn off the CHAT invitation so you won't be interrupted during the session. At Command Level, type CHAT –OFF <RETURN>. Got it? Now to go to the main menu, enter:

****

–> MENU <RETURN>

****

You should then see on your screen:

****

MENU OF SERVICES <MENU>

1 Today From The Source <TODAY>
2 News, Weather and Sports <NEWS>
3 Business and Investing <BUSINESS>
4 Communication Services <COMM>
5 Personal Computing <PC>
6 Travel Services <TRAVEL>
7 Shopping, Games and Leisure <HOME>
8 Member Information <INFO> **FREE**

Enter item number, <H>elp or <Q>uit:

****

As we mentioned, this top menu is a handy outline of the entire system—all The Source's features fall into one of these broad categories—so we can use it to figure out where we're going. On this tour, you'll be visiting two features in the Communication Services area (SourceMail and the Member Directory) and one in News, Weather and Sports (United Press International).

READING YOUR MAIL

The last time you looked at SourceMail, you wrote an electronic letter to yourself. Now, let's find it.

We'll be taking a couple of different routes to the post office in this session, but we'll start with the one you already know. Where the cursor has stopped at the bottom of the menu on your screen, type 4 <RETURN>, and you should see:

```
****************************************************************

COMMUNICATION SERVICES <COMM>

1 SourceMail
2 Bulletin Boards
3 Chatting Online
4 Computer Conferencing <PARTI>
5 Member Directory <MEMBERS>
6 Members Currently Online <ONLINE>
7 E-COM and Mailgram Messages
8 Member Publishing <PUBLIC>
9 How to Upload Files to The Source

Enter item number, <H>elp or <Q>uit:

****************************************************************
```

Now select Option 1, for:

```
****************************************************************

SOURCEMAIL

1 Overview
2 Instructions
3 Mail Check <MAILCK>
4 Begin Mail <MAIL>

Enter item number, <H>elp or <Q>uit:

****************************************************************
```

We'll be looking at the handy MAILCK feature in a minute, but for now, go directly into the SMail program by entering a 4 at the prompt, and you should see:

****

<S>end, <R>ead, <SC>an, <D>isplay, or <Q>uit?

****

Last time we skipped over all of these primary options except <S>end. There's nothing very mysterious about the rest of them. When you think about it, these are the main functions of *any* mail system, electronic or otherwise. You can write and mail letters (called Send), open and look at the letters you receive (Read), and browse the return addresses on the envelopes (SCan). And there are options for filing away letters you want to save and deleting letters you want to throw away.

Remember that whether you reach the SourceMail program through these slow, but friendly, menus or by the express lane of the Command Level prompt, you'll always be dealing with these simple functions.

The letter you wrote to yourself should be in your electronic mailbox by now, so try out the Read command. Type R <RETURN> at the prompt, and you should see something like this:

****

From: BBZ824 2-Lines
On:03 MAR 1985 At: 17:05
To: BBZ824
Subject: MY FIRST SMAIL

-More-

****

Press RETURN at the -More- prompt.

****

This is my very first SourceMail electronic letter
And I have addressed it to myself.

Disposition:

****

The "Disposition:" prompt at the end of the letter is The Source's way of asking what you'd like to do with this letter. You have quite a few options. To see precisely how many, you could type a question mark (?) at the prompt and the system would list each with a description.

We'll be looking at all the Disposition: options later. For now, all you need to know is that they break down into the same options you have with a paper letter you receive: you can leave it in your mailbox, reread it, reply to it immediately, forward it to someone else, save a copy in your filing cabinet, and/or throw it away.

Since you'll be using this letter again on this tour, just leave it in the mailbox. To do that, press RETURN at the Disposition: prompt (leaving the line blank). That tells SourceMail to leave the letter alone and continue.

Now the system will see if there's anything else in the mailbox and will display it if there is. (Since you're new to this neck of the woods, there probably isn't.) Eventually, The Source says, "No More Mail." You then get the option prompt again:

****

<S>end, <R>ead, <SC>an, <D>isplay, or <Q>uit?

****

Of course, the last option listed here takes you back to the Command Level, and that's just where we want to be. Enter Q <RETURN>.

SCANNING THINGS

Here at Command Level, let's try the Mail Check feature we were talking about. You could, of course, back up through the menus to reach the option, but instead let's use the command that was highlighted in the brackets on the menu we saw earlier. At the -> prompt, type MAILCK <RETURN>. After a hesitation, the system should report:

****

1 Read, 1 TOTAL

****

MAILCK is like sending an electronic servant down to the post office to see if there's anything worth your attention. In this case, he's reporting, "There's one letter in the mailbox, but you've already read it at least once." (If you hadn't already "opened" your letter, MAILCK would report "1 Unread, 1 TOTAL.") Many veteran Source subscribers make MAILCK the first order of business each time they log on.

As you've already figured out, no doubt, you can get to the SourceMail program without fussing with menus at all. To do that, you'd type MAIL <RETURN> at the Command Level prompt.

Do that now and the system says, "Do you want instructions?"

Do you? Typing Y here would display them, but in the interest of saving time, type N. We'll be covering the same ground, anyway.

Now The Source again shows you its compressed list of options:

****

<S>end, <R>ead, <SC>an, <D>isplay, or <Q>uit?

****

You've already seen how <S> and <R> work. <SC> stands for "scan" and will give you just the headers of the letters. Try that. Type:

****

SC <RETURN>

****

And what you should see is just the "return address" and the subject of your letter—with an important difference. This time the prompt at the end of the message says:

****

<R>ead or <D>elete by number:

****

Each message SMail scans is assigned a number. If you want to read the first message, you could enter R1, or if you want to delete it, enter D1. More about deleting things in a little while.

FASTER, FASTER

Can we do this mail operation even quicker? You bet. And this is an opportunity to learn about command strings. Enter a Q at the prompt on your screen to return to Command Level.

One of the convenient features of SourceMail—and The Source, in general, for that matter—is that commands can be strung together at the -> prompt. The more familiar you become with the system the faster you can travel by issuing several instructions in one line of commands.

Suppose you know that you want to run the MAIL program and SCan the messages in your mailbox. You could enter one command to do it all: MAIL SC.

Or instead of scanning, we could string together commands to Read everything in the mailbox. At the command prompt, type this—but don't press RETURN yet:

****

-> MAIL R (no RETURN yet)

****

Notice that this instruction really is two commands. We're telling The Source, "Run the MAIL program *and,* at the first option prompt, enter an R for Read."

Now, press RETURN.

After a moment's hesitation, the first thing you should notice is that the system has skipped right over that "Do you want instructions?" question. Smart system. It figures if you know enough to string commands together, you're sharp enough to get by without the instructions prompt.

Okay, go ahead and look at your message again (by pressing RETURN at the -More- prompt), and let's all meet again at the Disposition: prompt.

YOUR FILING CABINET AND YOUR MAILBOX

While we're waiting for our Disposition: to come around again, let's talk about the other option on the main prompt—<D>isplay. It's the command to show you any letters you have stored in your electronic filing cabinet.

And this leads us to an important distinction. On The Source (and in your home or office) you can have two different mail receptacles—a mailbox and a filing cabinet. They're different animals. You don't have your mail delivered to your filing cabinet, nor do you store your important messages permanently in your mailbox.

SMail considers your mailbox as a temporary holding area for messages. For permanent storage, you can have a filing cabinet.

To see how filing cabinets and the Display command work, we need to file something. So, at the Disposition: prompt now on your screen, type the word FILE <RETURN>. You're saying, "Put a copy of the letter I've just read in my filing cabinet."

Now, if you've never filed an SMail before, The Source probably will send you a message asking if you want to create a filing cabinet.

Here are the considerations:

—There are other ways to save your mail on-line, which we'll be showing you.

—It costs extra to have a filing cabinet—it requires 16 storage blocks, or about $6.20 a month at current rates, plus additional storage charges for letters you file there.

The advantage of a filing cabinet is that it gives you a handy way to save your electronic mail and even help you organize your correspondence. For instance, if you wanted to keep all your letters from George in a folder named GEORGE, you could type FILE GEORGE <RETURN>. If you don't specify a folder name with the FILE command, The Source will put it in a general-interest folder.

So, if you decide to have a filing cabinet, you can participate in the next few tour stops. If not, just read over the next few pages without actually typing the commands so you'll get an idea of how this feature works.

If you've established a filing cabinet and typed FILE <RETURN>, the system reports "Filed," then comes back to the Disposition: prompt. Press RETURN and you come to the "No More Mail" message, followed by the main prompt. After a moment, the system reports "Filed," and comes back to the Disposition: prompt. Press RETURN and you come to the "No More Mail" message, followed by the main prompt:

****

<S>end, <R>ead, <SC>an, <D>isplay, or <Q>uit?

****

Stop here and let's assess what we've found out after reading our mail for the umpteenth time.

1.We've met two major prompts. The one on the screen right now lists the main options. Some people find it useful to think of this as the "action prompt." The other, the Disposition: prompt which is displayed after you read a letter, lets you decide what you'd like to have done with the letter.

2. We've found out that you can take the express lane to the SourceMail program from the -> prompt of Command Level. MAILCK gives you a quick status report. Typing MAIL <RETURN> will take you to the SMail program, but you can save time by stringing commands together, such as MAIL SC <RETURN> or MAIL R <RETURN>.

3. You've seen that SourceMail actually has two kinds of storage: temporary storage in the mailbox and permanent storage in the filing cabinet. We've just placed a copy of your first letter in the cabinet by entering FILE at the Disposition: prompt.

This is where the Display command comes into play. At the prompt on your screen, type D <RETURN> and the system should show you the only other major prompt in the SMail service:

****

<F>iles or <D>istribution list?

****

Right now, don't worry about distribution lists. That leaves us something to talk about another time. Enter F, since you want to see what files you have stored, and The Source reports:

****

You have 1 letters in your filing cabinet
and 0 filing categories.

****

The "1 letters" is the one you just filed. The "0 filing categories" means that you haven't set up any special folders. Remember FILE GEORGE? That "filing categories" number would increase if you started putting letters into specific filing-cabinet folders.

Fine, but how do you read or scan the letters you've stashed away in the filing cabinet? "Display" *doesn't* display the mail at all—it's more of a status report, like MAILCK, except for the filing cabinet.

To look at what's been stored in the filing cabinet, you need to keep in mind what we're saying about the difference between the mailbox and the filing cabinet. You have to tell the MAIL program that you want to look into the filing cabinet—by adding the word FILE to your command.

Try this. At the prompt on your screen, type:

****

SC FILE <RETURN>

****

That command string tells the MAIL program to scan, not the mailbox, but the filing cabinet. If you enter the SC without the word FILE after it, the program will "assume" you mean the mailbox. In the same way, you can R FILE. Just remember that the magic word is FILE. With it, SMail will open the filing cabinet; without it, it will open the mailbox.

Is your mind racing ahead to the Command Level prompt? You're right. These filing-cabinet options can also be added to your command strings when you're traveling in the fast lane. For instance, at the -> prompt, you could enter MAIL SC FILE or MAIL R FILE to specify that you're talking about the filing cabinet, not the mailbox. Ah, you're going to learn to love these command strings because they can be used with virtually every service you'll be meeting on The Source.

HOUSECLEANING

The letter we've grown so tired of seeing is *still* in the mailbox. Remember that what we filed away was a *copy*. The original stays in the mailbox until we throw it away.

(There are several reasons The Source does that. Sometimes an account is used by more than one person in a family or an office. One person may discover a message meant for the other account user. Sometimes, too, you're just too busy to dispose of your new mail. You may want to read it now, but deal with it later.)

It's good that The Source gives you so much flexibility in managing your correspondence. However, you need to avoid letting your mailbox overflow. The system automatically deletes any messages left in the mailbox for more than thirty days, so you should routinely delete your letters after you've read them and decided on their disposition.

There are two ways to delete mail. You've already seen one—the SCan option automatically gives you an opportunity to delete messages by number. The other way is at the Disposition: prompt. After you've read and/or filed a message, you can simply enter D for Delete at the Disposition: prompt.

Deleting the mail from the mailbox does not affect the copy you've stored in the filing cabinet. When you are scanning or reading the stored messages in the filing cabinet (with SC FILE or R FILE), you'll have options to delete there, too.

THE WONDERS OF UFD

SourceMail has many more options and we'll be returning to the powerful SourceMail program in the next chapter. But, for now, the clock is ticking, so let's press on. We want to go to the place where this filed mail is stored—the User File Directory—so at the prompt on your screen, enter Q to get to Command Level.

Now, what do we have to do to find this UFD?

Nothing. We just did. The Command Level -> prompt also is the front door of your own personal storage area. The SMail you file automatically goes into a special section of this directory. You can even keep BASIC programs here, store files that you've received from other Source subscribers, and type in files directly. (Some subscribers keep form letters here for SMailing to a number of people.)

Let's type in a file. At the Command Level prompt, type this:

****

–> ENTER TEST <RETURN>

****

This is a kind of command string, like the ones you used in Source-Mail. The command here is "ENTER." The "TEST" part is the file name. So you're saying, "I'm ENTERing a file; you should call it TEST."

The Source replies; "Enter text."

Now type in the following text, pressing the RETURN key where we indicate. Don't worry about typos. We'll take up on-line editing in Chapter 6. All we want to do now is get a few files in the directory.

Here's the text:

****

This is the first file I've entered <RETURN>
directly into my file area, which, <RETURN>
on The Source, is called the User <RETURN>
File Directory, or UFD. <RETURN>
<RETURN>

****

When you pressed the second RETURN at the bottom of the file (that is, a RETURN on a new line), you closed the file and were returned to the –> prompt.

The file's closed, but where did it go? Let's see. At the Command Level prompt, enter LD (for List Directory) and your screen should look something like this:

****

-> LD
<USR12A>BBZ>BBZ824 <All access>
2 records in this directory, 37 total records out of quota of 1000
1 Files.
TEST
1 Directory.
MAIL.UFD

****

There's TEST at the top and there's something called MAIL.UFD. That's the filing cabinet you created for your SourceMail. (It's actually considered a "subdirectory.")

Create another file. At the Command Level prompt, type ENTER TEST2 <RETURN>.

By the way, about file names—we're using short ones. Actually, though, the names can be up to 32 characters. And you can use two words, if you'd like, but don't separate them with a space. Use a hyphen or a period. For instance, ENTER NEW.ORGANIZATION. You can't use numbers at the beginning of the file names, but it's acceptable to have them within or at the end of the name, as we've just done.

After the "Enter text" message comes from the system, type this:

****

This is the second file I've <RETURN>
entered into the UFD. <RETURN>
It's not particularly witty. <RETURN>
<RETURN>

****

Again, the second RETURN at the bottom closes the file.

Something else to remember: we've intentionally made the lines in these files very short. You can have up to 78 characters a line before tapping RETURN. It's a good idea, though, to get in the habit of pressing RETURN at the end of each screen line.

Now, here's a different kind of directory. Instead of LD, enter F (or FILES), and the screen should look something like this:

```
*****************************************************************

-> F
<USR12A>BBZ>BBZ824 (ALL), Records= 3, Quota= 24 / 1000

Files= 2.
ALL     1 sam                04 Mar 85 10:54:56 TEST2
ALL     1 sam                04 Mar 85 10:54:36 TEST

Directories= 1.
ALL     21 dir     dmp       04 Mar 85 10:10:32 MAIL.UFD

*****************************************************************
```

With this command, a lot more information is displayed about each file. You're told the creation date and time, the number of "records" (that is, an idea of the size of the files), the access rights (that's the "ALL" part—since you wrote them, you have all access rights).

All that's fine. However, as a practical matter, you usually won't need that much information about your directory. An LD command is faster.

Incidentally, some of this may be beginning to sound familiar to you, particularly if you're used to just about any modern microcomputer with a disk operating system. You'll find that many of the same disk file operations available with your DOS are available here. The commands probably will be different, but the functions will be very similar.

READING AND DELETING FILES

Before moving on to the last leg of the tour, let's see how to read files typed into your UFD, and finally, how to delete them.

The command to display a text file is TY (for TYpe). At the Command Level prompt, enter TY TEST (or TY TEST2) and the file will be displayed. No mystery there.

To delete, the command is DEL followed by the file name. (You're charged for what you store on-line as well as for your connect time, so you'll want to regularly review what's in your UFD to see if you want to keep it there.)

Also, you can save time by deleting several files at once. So, type: DEL TEST,TEST2 <RETURN> and The Source will respond:

****

```
TEST Deleted.
TEST2 Deleted.
->
```

****

By the way, don't try to manipulate your MAIL.UFD file here at the command prompt. It's best to use the SourceMail program itself for that.

THE SEARCHABLE SOURCE

We want to switch gears now and show you another part of the system. It will illustrate what many consider to be the single strongest feature of The Source: its search capabilities.

When you turn a computer loose on a mountain of data, it's awe-inspiring how quickly it can sort it and find material you've specified. The people at The Source have spent a lot of research time coming up with some remarkably fast searching utilities.

Before we call it a night, we'll check into two of them—United Press International and the Member Directory. Not only will they illustrate what The Source can do when you ask it to look up something, but we think you'll find both features fun to use.

First, news.

At the Command Level prompt, type UPI <RETURN>. After a few seconds, your screen should look something like this:

****

```
UPI DATANEWS
Type HELP or QUIT at any time.

<N>ational, <R>egional, <S>tate news or <ST>op:
```

****

You're now at the top of the UPI program, the same place you could have reached by selecting the second option from the main menu (News, Weather and Sports). Here, the system is asking which of its news wires you want to look at. The regional and state news options would next prompt you to enter a state code (such as NY for New York or CA for California). But, let's look at the national wire. Type N <RETURN>.

Now The Source says:

****

<G>eneral, <B>usiness, <S>ports, or <M>iscellaneous news:

****

All of the state, regional, and national wires are subdivided into these categories. (Hmmm—you're beginning to see the power of this thing: the same program that can give you general business news from the nation can provide sports news from Florida.)

Choose general. Enter a G, and the next option says:

****

Keywords (press return for all stories):

****

Here's where the fun starts. UPI is letting you look for all stories that mention a specific word or words. (As the prompt says, if you leave this line blank by just pressing RETURN, you can see *all* the stories on this wire. That would be fine for just catching up on the latest news. However, depending on the time of day and the wire you're scanning, it could mean hundreds or even thousands of stories. The route we're following will make UPI work a little harder for you.

Let's look for stories that mention a subject near to our hearts. At the "Keywords" prompt, type COMPUTER <RETURN>. (It doesn't have to be all capital letters, of course.) The Source says:

****

Enter starting & ending date—or press return for today:

****

Like many of The Source's reference services, UPI keeps a backlog of material. In this case, seven days' worth of wire stories are on-line for your reading pleasure, and you can be very specific about the time period you're interested in. You can ask for stories between two dates, before a certain date, since a certain date, or, as the prompt tells you, just those filed today.

Enter the dates in numeric form. Suppose it was March 7. You could ask to search stories between March 2 and March 5 by entering 3/2-3/5. For stories prior to March 6, you'd enter -3/6, and for those since March 3, you'd enter 3/3-, etc.

Let's give it a workout. Subtract, oh, six days from the current date and enter that at the prompt, like 3/1- <RETURN>. The system reports:

****

Searching.

****

Then

****

Pick a starting story number from 1
(the earliest) to 42 (the latest):

****

Pretty fast, wot?

Obviously, the high number will be different on your screen, depending on how many stories UPI found for you. When we did it, there were forty-two stories on the national general news wire in the previous six days that mentioned the word "computer" somewhere in them.

The system is ready to show them to you, but it wants to know where to start—with the earliest, the latest, or somewhere in the middle? We'd like the latest (i.e., most recent) stories first, so enter the larger number. (In our example, it would be 42.)

The next prompt you receive is this:

****

Read forward in time (RF), read backward
(RB), scan forward (SF), or scan backward (SB):

****

The Source is essentially asking two things: (1) do you want to read them or scan them, and (2) do you want to go forward or backward in time?

For us, forty-two is far too many stories to read. Therefore, we want to scan them in order to pick out the ones that look interesting. And, since we've already indicated at the previous prompt that we would like to start our list with the latest stories, we want to scan them backward in time.

Type SB <RETURN> at that prompt, for Scan Backward.

Finally, UPI is ready to get down to business. You've answered all the queries, so it will begin listing the stories it found, something like this:

****

42 03-07-85 08:26 pes

(updating-adding info)
UPI News at a Glance
By United Press International
WASHINGTON _ In a dramatic outburst,
Sen. Joseph Biden, D-Del., told Edwin Meese
Wednesday he will vote against his confirmation

41 03-07-85 05:29 pes
By SYDNEY SHAW
WASHINGTON (UPI) _ The Federal
Communications Commission, re-examining its
landmark 1980 rule that deregulated telephone

****

Etc.

Remember, you can use CONTROL S to freeze the display if it's coming in faster than you can read it. CONTROL Q restarts it.

Note that you're given the date and time each story was filed, the first few lines so you can see if it might be interesting, and a number with which to retrieve the whole story. (The Read Forward or Backward options would have displayed the entire stories, of course, instead of just the first few lines.)

Do you have to look at this whole list (forty-two stories!) before telling the system what you want to read?

Not at all. If you see a story that looks interesting, send The Source the BREAK signal—CONTROL P will do it. Try that. As the scan is continuing on your screen, type CONTROL P.

After a hesitation, you should receive a message like this:

****

Break. . .

We are on story 40.
Type R, S, N, B, or G and a Story Number:

****

That means that by the time the brakes were applied, we reached story number 40, and the system is waiting for further instructions. The letters in the last line are hints of what options are open.

The "G" here means Get. Suppose that story 41 looked interesting. You could enter G 41 to "get story 41."

Give that a whirl. Pick a story on your list and get it.

But what if you read a few paragraphs of the story and decide it's not as interesting as you thought it would be? Send another BREAK signal—CONTROL P—and in a few seconds (allowing for the BREAK to catch up with the system), you should see:

****

Break. . .
We are on story 41.
Type R, S, N, B, or G and a Story Number:

****

Now to get our scan going again, use either the B or the N option mentioned in the prompt. They mean Back and Next and each can have a number with it. If you type N 2 <RETURN>, you're saying, "Jump forward to the *next* two stories from where we are now (story 41, in our example) and resume the scan." B 2 would mean to back up two stories and resume. The R option Restarts the scan where you left off. The S option will Stop the scan and return you to a prompt that lets you select a different news wire.

Not listed in the options is P (the letter, not the CONTROL P). As you found out on the last tour, P means Previous and it will work just about anywhere on The Source to take you to the previous level of commands.

Try that—enter P at the prompt and you'll be whisked back to:

****

<N>ational, <R>egional, <S>tate news or <ST>op:

****

Also not shown here is Q. The QUIT command will work here as well as anywhere in the system; use it if you want to leave the UPI program and return to the Command Level.

We hope we've whetted your appetite for news sufficiently that you'll return to UPI for a little extracurricular exploration. Or perhaps you'd like to experiment with it right now. We can wait.

MORE TIPS

While you're doing that, we have a few more words about UPI.

First, about keywords. You probably noticed back there that the prompt for keywords was plural. We entered only one word, COMPUTER, but you can enter up to three, separated by AND or OR. For instance, if you wanted to look for stories that were about Reagan's tax proposals, you could enter REAGAN AND TAX. On the other hand, IBM OR TANDY would find stories that mentioned one or the other of the two companies, but not necessarily both.

Now, about taking the express lane. We got to UPI faster than we would have had we gone by way of the menus. However, by using some clever command stringing, we could have done it even faster. The same rules apply here as in command stringing in SourceMail—you simply string together answers to the prompts you know the system will be looking for.

You might want to review these past few pages to see how you could do that. For instance, if you wanted to search and "Scan Backward" UPI's business wires for stories about "Honeywell" over the past several days, you could enter most of the strategy in one neat line at the –> prompt. You'd just need to be carefuly to string the commands in the order in which the system is expecting them:

****

–> UPI N B HONEYWELL 3/1– SB

****

Of course, the date (3/1- here) would be different, but you get the idea. The "UPI" part of the string would take you to the UPI feature; "N" would choose the national news category; "B" would select the business wire; 3/1– would be the date specification; and "SB" would anticipate the final prompt by saying Scan Backward.

Naturally, you wouldn't have to enter the entire strategy at one time. If you wanted, you could type just UPI N B <RETURN> to take you up to the Keywords prompt.

FINDING FELLOW TELETRAVELERS

Leave UPI and return to the Command Level prompt. (Enter Q.)

Our last on-line stop—the Member Directory—shows you even more powerful searching tools available on The Source. To get to it, type MEMBERS <RETURN> at the –> prompt, and after a moment, The Source will show you:

****

THE SOURCE Member Directory.

1 Search the Directory
2 Add Your Entry
3 Modify Your Entry
4 Replace Your Entry
5 Delete Your Entry

<I>nstructions
Enter Item Number or <H>elp:

****

In a minute we'll suggest you file your own biographical material, but first, let's see what's happening here. Enter 1 to Search the Directory, and the system responds with:

****

Search by:
<ME>mber ID <NA>me <CI>ty <ST>ate <CO>untry
<O>ccupation <MI>cro <IN>terest or <ALL>:

****

Like UPI, the directory program starts out wanting to know which area of each user's record you want to search. (People who do a lot of database searching sometimes call these topics *fields*. Think of them as column headings on an index card or ledger book.)

We'll search first by the "microcomputer" field. Enter MI, as the prompt says, and The Source displays:

****

Enter Micro:

****

We can find other Source subscribers who use IBMs. Type IBM <RETURN> at the prompt.

****

Searching. . .
2066 items found.

<N>arrow <E>xpand <SC>an <ME>nu <C>ancel:

****

Well, the good news is, it sure didn't take The Source long to locate more than 2,000 subscribers who say they use IBMs. The bad news is, it would take a heck of a long time to look over that many records.

Fortunately, this service, unlike the UPI feature, lets you "<N>arrow" or "<E>xpand" the group you caught in your net in the first search.

At the prompt, enter N for Narrow, and the system gives you this prompt again:

```
****************************************************************

Search by:
<ME>mber ID <NA>me <CI>ty <ST>ate <CO>untry
<O>ccupation <MI>cro <IN>terest or
<ALL>:

****************************************************************
```

Let's see how many of those IBM users say they're interested in programming. Select the "<IN>terest" field (enter IN), and after the system displays

```
****************************************************************

Interest, or <H>elp:

****************************************************************
```

type PROGRAM <RETURN>. (That'll find those who list interests such as PROGRAMMER, WRITING PROGRAMS, ORIGINAL PROGRAMMING, etc.)

```
****************************************************************

Searching. . .
134 items found.

<N>arrow <E>xpand <SC>an <ME>nu <C>ancel:

****************************************************************
```

That chopped down our 2,066 records considerably. However, that's still too many for a casual scan, so let's Narrow the group further. Select N again. (Incidentally, Expand works like Narrow, except in the other direction. It will let you add records to your latest group by defining another search.)

Right now, though, we're narrowing. How about finding out how many of those IBM programmers live in California? At the prompt,

****

Search by:
<ME>mber ID <NA>me <CI>ty <ST>ate <CO>untry
<O>ccupation <MI>cro <IN>terest or
<ALL>:

****

enter ST for STate, and when you're next asked for "State, or <L>ist:," enter CA for "California." (As the prompt says, L will give you a list of recognized state abbreviations. It generally follows the U.S. postal codes for the states.)

****

Searching. . .
23 items found.

<N>arrow <E>xpand <SC>an <ME>nu <C>ancel:

****

Ah-hah—twenty-three IBM programming enthusiasts from California. That's a more manageable list. At this point, you can either SCan (that will show you the complete entry on each—name, user ID, location, interests, etc.), or you can choose ME and The Source will build a menu for you, using the first line of each record it's found. C, of course, will cancel the search.

Why not look at a few of them? If you choose to SCan, you'll be shown the first few, then the screen will pause after sending a "Press return for more:" prompt. Tap RETURN to get the next few; P (for

Previous) will break out of the search and take you back to the previous option prompt.

After you've looked at some of your list, enter P and get back to the top menu of this service:

****

```
1 Search the Directory
2 Add Your Entry
3 Modify Your Entry
4 Replace Your Entry
5 Delete Your Entry

<I>nstructions
Enter Item Number or <H>elp:
```

****

The Source Member Directory is a valuable way to find users with similar interests. Regulars on the system use it to reach out to people via SourceMail, inviting them to carry on a correspondence, maybe meet for a CHAT sometime.

MAKING YOUR MARK

Entry in the Member Directory is voluntary, so it doesn't list *every* subscriber—only those who have chosen to be listed. Since there's no extra charge to be in the directory, we suggest you select Option 2 here and enter your own record. You'll be prompted for your name. (Should you use your full name? That's your decision. Many people do, but a first name only is certainly acceptable.)

You'll also have an opportunity to enter the name of your city and state, your occupation and interests, and the kind of computer you use—all the kinds of things you'd probably want to know about somebody else on-line. Of course, don't list anything that gives away your password. And you don't have to fill in all the blanks. If you want to skip a question, just tap RETURN.

We'll leave you here so you can do a few more searches. Perhaps you'd like to find other Source subscribers in your state, others who might be interested in sports, chess, cooking, Spanish folk songs, whatever.

When you're finished, enter Q at a prompt to get back to Command Level. Then enter OFF to disconnect. Save the last part of this chapter for perusal after you've logged off.

MORE ABOUT SOURCEMAIL

In this session you went around the block in the SourceMail program. You read your first letter, scanned it, saved it in the filing cabinet, and deleted it from the mailbox. In fact, with a few more commands, you'll know everything there is to know about how to be on the receiving end of SMail. Then in a later session, we'll fill in some gaps on how to write and send mail.

Some of SourceMail's greatest power comes from the flexibility it gives you to manage your mail. In this chapter you saw four of the major commands you can enter at the Disposition: prompt that comes after you've read a letter. You can enter:

<RETURN> (leaving the Disposition: line blank) to leave it in the mailbox and continue to the next message.

D to Delete the message you've just read.

Q or QUIT to leave the message in the mailbox, leave the SourceMail program altogether, and go to the -> prompt of Command Level.

FILE to put a *copy* (not the original) of the letter in your electronic filing cabinet. In addition, you can enter FILE followed by a category name to store the message in a "file folder" in the filing cabinet.

Besides these, there are four other major commands you haven't met yet. At the Disposition: prompt, you can also enter the following:

AG or AGAIN displays the message again.

RE or REPLY sends an answer to the person who wrote to you. The system will take care of remembering the user ID for you. And, as you learned earlier, entering .S (that is, a period and an S) will send the reply after you've finished it.

FO or FORWARD (followed by a user ID number) sends a copy of the letter you've received to another Source member.

SAVE (followed by a file name) keeps the letter, not in the filing cabinet, but as a separate file in your User File Directory. It would be just as if you had typed the message directly into your UFD. You can also enter SAVE IN (and a file name) to append the letter to an existing file in your User File Directory. But don't confuse these with the

FILE commands. SAVE and SAVE IN put the message in the main User File Directory, while FILE puts it in SourceMail's filing cabinet.

You've seen that reading and scanning messages can be done from menus and short prompts. But to save time, you can also link commands together, such as R FILE or SC FILE. And you can even do it from the -> prompt with a command string that starts by summoning the SMail program.

There are a number of other options to add to those command strings. You saw one on this tour. R FILE or SC FILE will Read or SCan letters you've filed in your filing cabinet sometime earlier. You can also specify a particular file folder by adding a category after the FILE command. For instance, R FILE SALES would show you all the letters you FILEd in the SALES folder.

There are a few more possible additions to a Read or a SCan command. For instance:

UNREAD or UN (as in R UN or SC UN) reads or scans new letters—those you haven't read before.

FR followed by a user ID (R FR BBP112 or SC FR BBP112) reads or scans messages from someone with a specified user ID.

DA (followed by the day, month, and year, as 15 JAN 1985) reads or scans mail sent on a specific date.

SU (followed by a keyword) reads or scans those that contain a certain keyword in their Subject: line.

Other commands for the SourceMail program have been summarized in the On-line Survival Kit at the back of the book.

SUMMING UP

In this chapter, in addition to some major SourceMail features, you learned these commands that can be entered at the -> prompt of Command Level:

MAILCK which gives you a quick status report on your mailbox, telling you how many messages are there and how many are unread.

ENTER, followed by a file name, which lets you write a file in your User File Directory. After you've written the text, you can close and store the file by tapping RETURN on a new line.

TY, followed by a file name, which displays (or TYpes) that file after it's been closed.

LD (for List Directory) which gives you a brief catalogue of the files stored in your UFD. F or FILES will give you a more detailed catalogue.

DEL, followed by a file name, which deletes that file. And you can delete more than one file by stringing them together with commas, such as DEL TEST1, TEST2.

CHAPTER 6

Is There an Editor in the House?

The Source is an interactive medium and you won't get your fair share of what it has to offer unless you can communicate with the people you'll meet there. That means learning to write and edit on-line.

You've already written a few things on the system—an electronic letter to yourself, a couple of items in your private file area (UFD), perhaps an entry in the Member Directory. But so far, we've steered you away from some of The Source's most exciting features because they require you to share your written thoughts with others.

To rectify that situation, we need to spend the next few minutes introducing you to The Source's on-line editing tools.

Uh, yes, we hear the loud yawning. It's inevitable that some people think learning about electronic writing and editing is as boring as freshman comp. We think you'll be pleasantly surprised, though.

For one thing, the Source's editing program is logical and pretty easy to learn. More importantly, the EDITOR is the key to a whole new level of systems "tricks" that can make navigating the system faster, easier, and cheaper. In this chapter, you'll see how a little basic editing knowledge can help you construct what's called a "command file" that lets you get more mileage out of the command strings you've been learning. And that's just the beginning.

Come on-line with us now and we'll show you that writing and editing on The Source is more than just commas and colons.

ON THE LINE AGAIN . . .

Fire up the computer and log on to The Source as usual. At the first menu, go to the Command Level.

Usually at this point, we ask you to enter CHAT -OFF (with a space between CHAT and the hyphen) to automatically turn down CHAT requests so we won't be disturbed. But this time, let's try something different. At the Command Level prompt, type:

****

-> ENTER SHH <RETURN>

****

The system will open a new file, call it SHH, and signal us to "Enter text." When it does, type this:

****

CHAT -OFF <RETURN>
<RETURN>

****

The second RETURN has closed the file and returned you to the -> prompt. Now type:

****

-> CO SHH <RETURN>

****

After you press RETURN, you should see The Source display the words "CHAT -OFF" before taking you back to the -> prompt. This new command—CO—causes the system to execute whatever command is contained in the file you specify. Had there been other command strings in the SHH file, they would have been executed, too, one after another.

Hmmm! Some rich possibilities in these new command files. Hold that thought. We're coming back to this in a few minutes. However, in order to add commands to this file, we need to know more about editing.

EDITING ON-LINE

To do that, let's create a new text file to work with. At the –> prompt, open a file called STUFF, by typing:

****

```
-> ENTER STUFF <RETURN>
```

****

After the system says "Enter text," please type these few lines, tapping the spacebar four or five times at the beginning of each paragraph. Be sure to put in the typographical errors we've so carefully added. Watch for "telll," "persauding," "traffice," "laywers," and "wrods."

****

```
     This is where Bowen and Peyton telll us <RETURN>
how to make an editor do what we want. Hmmm, <RETURN>
wonder if they have any tips on persauding <RETURN>
traffice cops and laywers as well? <RETURN>
     Down to business—this file was written <RETURN>
into the User File Directory. We just typed the <RETURN>
wrods ENTER STUFF at the Command Level prompt <RETURN>
and began typing, remembering to tap the <RETURN>
ENTER or RETURN key at least every 78 characters. <RETURN>
     To close the file, we tap RETURN on a <RETURN>
new line. <RETURN>
<RETURN>
```

****

Pressing RETURN on a fresh line closed the file and brought you back to the –> prompt. Nothing new so far. But what do we do about all those typos? To tell The Source you'd like to edit this file, type:

****

```
-> ED STUFF <RETURN>
```

****

You will receive a message that says "EDIT."

Here's what's happened: the system has called on its EDITOR program (it did that when it saw the ED part of the command) and called up the STUFF file. It's now in EDIT mode, as the message indicated. There's another mode, INPUT, which we'll be seeing in a minute.

But where's the file itself?

To figure that out, you need to know something about the way the EDITOR program operates and how it probably differs from your word processor. Start by thinking in terms of an "invisible pointer" that the EDITOR uses to move up and down a document. Right now the pointer is *above* the first line of your text (that is, above the line that begins, "This is where Bowen and Peyton telll us").

We can see that by printing the first few lines. Type:

****

```
P2 <RETURN>
```

****

(This means "print two lines.") The system should display:

****

```
.NULL.
      This is where Bowen and Peyton telll us
```

****

Ah-hah. This first line was ".NULL," or nothing, and that's where the pointer was when we started. The next is "This is where Bowen and Peyton . . . ," etc. Let's try that again. This time enter P3 <RETURN> and The Source shows you three lines—the current one and the next two:

****

This is where Bowen and Peyton telll us
how to make an editor do what we want. Hmmm,
wonder if they have any tips on persuading

****

An important note: as it prints the lines in response to the P command, the EDITOR also is moving its invisible pointer down the file, one line at a time. The pointer started out above the first line (at the .NULL), and it has moved down to the line that starts "wonder if they . . ." The present location of the pointer is sometimes thought of as the "current line." (Confused? Don't worry—you'll get the hang of it as we play with this moving finger.)

To move the pointer back to the top of the document, type T <RETURN> (that means Top). Now type P2 <RETURN> again and see:

****

.NULL.
This is where Bowen and Peyton telll us

****

So, here we are again. The T took us back topside, and the P2 showed us the first line (the .NULL) and then the first line you typed in, "This is where . . ." And in the process, the EDITOR has moved the pointer to that line.

SWATTING TYPOS

The typographical mistakes we asked you to include are beginning to show up. The first is in the line now on your screen ("telll"). Fixing it means changing *lll* to *ll*—and Change is the command we'll use. Type

****

C/telll/tell <RETURN>

****

(with no spaces). This translates, of course, to "change telll to tell." Note how you're using those slashes (/) to offset the portion of the line you're changing. The old section of the line (or "string") goes first, right after the C/.

After you enter the command, the system should display the changed line:

****

This is where Bowen and Peyton tell us

****

Three notes:

First, you really don't have to type in the entire word you want to change—just the unique part of the word or line will do the trick. For example, "C/lll/ll" would do it, meaning, "Change those three *l*'s to two *l*'s."

Second, we're using uppercase in all our editing commands, but that's just to help you spot them in the examples. The commands here, as virtually anywhere in The Source, can be either upper- or lowercase.

Finally, and most important, the reason this change worked was because the pointer was on the line where the word *telll* occurred. Remember, to fix an error in another line, you first have to move the pointer to that line.

And that's what we need to do now. As you may recall, we gave you a misspelled *persuading* (as *persauding*) when you typed in this little ditty. We can fix that one next. But how do you know where to move the line pointer?

Well, you *could* print each line, entering P2 <RETURN> to display the current line and the next one, and keep entering P2's until you finally found it. On a short file like this, that wouldn't take too long. Then again, the file could have hundreds of lines.

Fortunately, the EDITOR has some better functions. One will locate the next line that contains a specified string of characters. Let's go back to the Top (enter T) and then type:

****

L per <RETURN>

****

Be careful that there's no space after the *r* in *per*. That told The Source, "Scan down the document from the current line position (which is now the Top) and find the first occurrence of *per*." (Had you put a space after *per*, the EDITOR, taking you literally, would have searched for *per* <space>.)

After a second or so, the system should display:

****

wonder if they have any tips on persauding

****

Okay, but why did we type the T command first?

The Locate (L) command—like all the find commands you'll be meeting—always searches *below* the current position of the pointer. So if you want to search the entire document for a string, you should first move the pointer back to the Top.

To fix the error in this line, you could type:

****

C/sau/sua <RETURN>

****

Or:

****

C/persau/persua <RETURN>

****

Take your pick. Either will work, and we're sure you see other variations. All that matters is that you get the *sau* part of the line changed to *sua*.

Now, flip back a page or so and look at the text of the message to see where our next error occurs. It should be in the line that begins "traffice cops and . . ." To move the pointer to that line on your screen, try:

****

L cops <RETURN>

****

This time we didn't bother to return to the Top, because we happened to know that the next typo is below the current position of the pointer.

Now the system shows us the line:

****

traffice cops and laywers as well?

****

We have two errors here. First, we need to whack that *e* from the end of *traffic*.

To the EDITOR, spaces are just like any other characters, so we can tell it to "change e-space to just space." Converted to EDITOR language, that's:

****

C/e / <RETURN>

****

(Note that there's a space after the second slash.) Now the system shows us the changed line.

****

traffic cops and laywers as well?

****

We still have a problem with *laywers*. Since we haven't moved the line pointer, you can just type:

****

C/yw/wy <RETURN>

****

It looks as if you're getting the hang of it. Let's do another Locate. Our next error is *wrods* instead of *words,* so type

****

L wr <RETURN>

****

leaving no space after the *r*. The system says—oops! Wait a minute. . . . What's this? It's printed:

****

Down to business—this file was written

****

That's not the line we're looking for; *wrods* isn't in there.

Ah, but the EDITOR did precisely what we asked. It located the *next* occurrence of *wr,* which happened to be the *wr* in *written*. The moral here is we'll find errors faster if we're more specific with the Locate command. Try

****

L wrod <RETURN>

****

(with no space after the d) and up comes

****

wrods ENTER STUFF at the Command Level prompt

****

Fix that line with the Change command and we'll see how to save your editing changes.

Ready? Type the word FILE <RETURN>. After a hesitation, the system reports the name of the file (STUFF) and then returns you to the -> prompt.

NAVIGATING YOUR TEXT FILES

To read your file again, use the command you learned the last time we passed this way—TY (for TYpe). In this case, you'd enter TY STUFF <RETURN> to see the latest version of your text, with the changes you've made.

It's becoming apparent, we're sure, that The Source's EDITOR considers your text as a collection of individual lines. That's why P2, P3, P12, etc., work so conveniently.

You can also have the program take you directly to a particular line—in other words, "Move the pointer to line 5, but don't bother to display the lines as you go."

Open the file for editing again by typing ED STUFF <RETURN> and, after the EDIT message appears, type:

****

PO5 <RETURN>

****

(that's a letter *O* rather than a zero) and The Source says:

****

Down to business—this file was written

****

PO means POint, as in "Point the invisible pointer at line 5."

Fine. But what if you don't know the number of the line you're looking for?

Well, a handy editing command is N, meaning Next line. It simply moves the pointer down a line. Try it. Type N <RETURN> and the system shows you the next line—"into the User File Directory . . ."

The nice thing about the Next command is that it can be used with a number—a positive number to go down so many lines, and a negative number to back up so many lines.

Let's play with that. Type N-1 <RETURN>. The Source will back up a line and show you, "Down to business—this file was written."

Now type N-3 <RETURN>. The system moves back three more lines—"how to make an editor do . . ."

Now try N3 <RETURN> (that is, 3 rather than -3). It goes down again (positive numbers mean forward, negative numbers mean backward) and shows you, "Down to business—this file . . ."

Obviously, if you found any errors as you moved about, you could fix them with the Change command, since the pointer would be aimed at the line. In other words, it doesn't matter *how* you get the pointer there—with a P for Print, an N for Next, an L for Locate, a PO for POint.

Before we move on, there's one more command you need to know—the F (for Find) command. It's similar to the Locate command, but more specialized. It looks only at the first part of each line and will find the line that begins with a word or letter you've specified.

To try out Find, go back to the Top of the text with the T command, then type

****

F in<RETURN>

****

(leaving no space after *n*). That means, "Find the line that begins with *in*." It should display "into the User File Directory. We . . ." Note that if you had used the Locate command—L in <RETURN>—the pointer would have stopped on an earlier line (line 3, which contains *persuading*.) The F command looks only at the first part of the line.

NUMBERS, PLEASE!

Sometimes it's useful to know the number of the line at which the invisible finger points. For that, the EDITOR has a Where command.

Type W <RETURN>. You should now be on

****

LINE 6

****

as the Where command just reported.

Some Source users like to have the EDITOR show them all the line numbers while they edit. Try this. Type:

****

MODE NUMBER <RETURN>

****

Nothing appears to happen. However, when you go back to the Top of the file (enter T), and now type a Print command, say, P999 <RETURN>, you see:

```
******************************************************************
.NULL.
00001:      This is where Bowen and Peyton tell us
00002:how to make an editor do what we want. Hmmm,
00003:wonder if they have any tips on persuading
00004:traffic cops and lawyers as well?
00005:      Down to business—this file was written
00006:into the User File Directory. We just typed the
00007:words ENTER STUFF at the Command Level prompt
00008:and began typing, remember to tap the
00009:ENTER or RETURN key at least every 78 characters.
00010:      To close the file, we tap RETURN on a
00011:new line.
BOTTOM
******************************************************************
```

Of course, these numbers are just an editing aid, they have not really gone into your text. Think of them as blue lines on your copy paper. MODE NUMBER is available just to show you where everything is.

SPEAKING OF MODES

Getting tired of STUFF? Hang in there—the grand finale is coming up. Then we'll apply some of this knowledge to your new CO file.

We said that the EDITOR works in two "modes": EDIT and INPUT. The program *toggles* between the two modes, just as a light switch toggles between on and off.

Look at the difference. Just tap the RETURN key on a line by itself and the system should respond:

```
******************************************************************
INPUT
******************************************************************
```

Tap RETURN again:

****

EDIT

****

Of the two modes, EDIT is the workaholic—it handles all the changes you make, it deals with locating the material and moving the pointer around. On the other hand, INPUT is interested *only* in new text.

Let's give that a test run. You should be in EDIT mode now. (If not, press RETURN again to get to EDIT mode.)

Ready? Type B <RETURN> to move the pointer to the Bottom of the text. Now we'll add a new paragraph.

Tap the RETURN key to toggle into INPUT mode. Now type:

****

This is a brand-new paragraph, added to <RETURN>
the bottom of the file. <RETURN>

****

At this point, you're still in INPUT mode, on a new line *after* "the bottom of the file."

Press RETURN and the system says, "EDIT," meaning we've left INPUT and returned to the original EDIT mode.

Now if you enter P <RETURN> to print the current line, you should see:

****

00013:the bottom of the file.

****

You'll always have the two modes straight if you keep in mind that EDIT is the responsible, hardworking member of the family, who handles moving the invisible pointer around, making all the changes you need, locating (or finding) the lines and line numbers, and carrying on the family name—EDITOR. INPUT has a limited attention span. It's only interested in new material. And that's why you don't call this an INPUTER.

MAKING GLOBAL CHANGES

One more editing trick before we retire the STUFF file for good. Let's get into EDIT mode to see how to make global changes in a text file.

Probably the word processor on your microcomputer has a "global search-replace." It can find all references to a word or phrase and change them to a different word or phrase. This EDITOR can do that, too. Start at the Top of the text (enter T), then type:

****

C/the /THE /G999 <RETURN>

****

The translation is, "Change *the* to *THE* and make the changes in the next 999 lines." (G, of course, stands for Global.) Since our text has only a dozen or so lines, the 999 will mean, "Do it till you come to the end of the file."

After this command, if you go back to the top of the file and enter P999 <RETURN>, you'll find that every *the* has become *THE*.

A couple of things about global changes:

1. It works on all of the lines below the one being pointed to at the moment. So, if you want to make the changes in the entire document, be sure to enter a T command first to get back to the Top.
2. You can put any number you like after the G—if you put G10 in the command string, it will search and replace for the next ten lines.

CLEANING UP

If STUFF were something of importance and you wanted to save it in its new edited state, you'd enter FILE right now and the system would delete the old version and keep the latest. However, since we're going to throw this away, anyway, try something else—type Q or QUIT <RETURN>.

If you're in EDIT mode, as you should be, The Source will report, "FILE MODIFIED, OK TO QUIT?" It's asking for confirmation that you want to throw away all the editing changes you've made. Type Y <RETURN> and you'll be back at the -> prompt.

Go ahead now and delete that file with DEL STUFF <RETURN> at the -> prompt.

PUTTING THE EDITOR TO WORK

Now that we've had an excursion into editing, you're almost ready to create any document on-line. (We'll fill a few gaps in the information after we've disconnected from the system.)

Also, we now have enough information to modify SHH, the CO file we're building.

We noted that CO files can, and usually do, contain more than one command string. They're The Source's way of storing up energy, because they let you get the most out of the fewest possible keystrokes.

With what we know so far, we can design a CO file that will:

1. Turn off CHAT.
2. Check the mailbox.
3. Look up today's computer news on the UPI wire.

We already have the first step in SHH. We put in CHAT -OFF when we started. Now let's edit it to add the other two. At the -> prompt, type ED SHH <RETURN>.

To display what we have so far, type P999 <RETURN>, and you'll see:

```
****************************************************************

.NULL.
CHAT -OFF
BOTTOM

****************************************************************
```

Printing it also moved the pointer to the bottom of the file. (We could have done the same thing by entering the B command.)

Now we're ready to add some new lines by getting into INPUT mode, so press RETURN. When you see the "INPUT" message from the system, type in the command for step two:

****

MAILCK <RETURN>

****

The third step—searching the UPI wires—is another one we learned in the last chapter. Type this string:

****

UPI N G COMPUTER <RETURN>

****

That's all we want to add for the time being, so we need to close the file. Press RETURN to get back into EDIT mode, then type FILE <RETURN> to replace the old version of SHH with this altered one.

When you've received the -> prompt again, let's see the new, improved SHH do its stuff. Type:

****

-> CO SHH <RETURN>

****

Now just sit back and watch the show. By itself, the system will display and execute CHAT -OFF, followed by the MAILCK command. After that, it will head for the UPI service, select the <N>ational, <G>eneral news wire, define the keyword as "COMPUTER," and finally come to rest at the prompt asking for the date at which you wish to begin your search.

Ahem.

A little display of power, isn't it? It'll get even better in the next chapter when we make some interesting modifications to this file.

If you'd like, go ahead and enter a date and continue the search for computer news. If not, enter Q at the prompt to return to the Command Level.

Either way, when you're finished, log off and make yourself comfortable. We'll close by rounding out the discussion of editing and seeing how it applies to SourceMail.

ADDING TO THE TOOLBOX—DELETE, INSERT, REPLACE

Now that we've turned off the meter, let's add a few more editing tools. You don't need to memorize all these. Like so many things about computers, the commands will become more familiar to you as you begin using them regularly. For now, it's enough just knowing they're available. All of them will work whether what you're writing is a letter or a command file.

Sometimes you may want to delete several lines from the file you're editing. Suppose in STUFF you'd noticed that five lines in a row were unnecessary.

Here's how you could remove them: locate the pointer on the first line of the series, using one of the navigation commands; then type D5 <RETURN> to delete five lines.

Or you could delete one line at a time by typing D <RETURN> (with no number). Then type P <RETURN> to see the next line. If you want it deleted, too, enter D <RETURN> again, and so on. Remember that the EDITOR starts deleting on the current line, so D5 actually means, "Delete the current line and the next four."

The EDITOR also gives you an easy way to replace a line of text. Start by locating the line you want to replace (with the L command, perhaps), then type R followed by the new text.

For instance, you might type L/this text <RETURN> to find:

****

```
I should replace this text
```

****

Then you could type:

****

```
R I have replaced this line<RETURN>
```

****

Actually, you won't use the Replace command very often in ordinary text files. You're more likely to change individual words and phrases (with the C/old word/new word) rather than entire lines. However, the R command is very valuable in editing the command (CO) files. There, you very well might want to replace the command string in one line with another command string.

The EDITOR also has several good ways to insert information into files. If you're inserting a single line, it's easier to do it in EDIT mode: locate the line below which you want to insert text, then enter I <space> and the text to be inserted (not to exceed 78 characters, of course). Such as:

****

I this is inserted material. <RETURN>

****

The Insert command will put the new line *below* the line being pointed to.

If you're planning to insert a number of lines, it's easier to toggle into INPUT mode. Locate the line (again, it's the line *below* which you want the material added), and press the RETURN key to move from EDIT to INPUT mode. Then type away. (Of course, as everywhere on The Source, you should press RETURN before the seventy-eighth character on each line.) Then toggle back to EDIT mode when you're done.

And here's a very helpful command, speaking of inserting. You can copy material from one file into a new one by using the LOAD command. To do that, move the pointer to the line *below* which you want the information added and enter LOAD followed by the file name. Suppose you had a file of addresses called ADDS that you wanted to insert into the new file you're working on called NEWGUY. While editing NEWGUY, you could type:

****

LOAD ADDS <RETURN>

****

The system would pick up a copy of the ADDS file (leaving the original intact, by the way), insert it into NEWGUY, and let you continue writing.

STORING FORM LETTERS FOR SOURCEMAIL

Ah-hah—we're getting close to what we were saying about storing form letters on-line, aren't we? Some powerful possibilities come to light when the EDITOR and the SourceMail program work together.

As you've already seen, anything you receive through SourceMail can be stored directly in a file in your User File Directory. (To refresh your memory on that, you might want to review the comments on the SAVE and SAVE IN commands at the end of the last chapter.)

It works both ways. Any text files you have on storage in the UFD can be mailed through SourceMail, using a variation of the LOAD command we just learned.

Suppose, for instance, that the NEWGUY file contained some general biographical material about you—a couple of paragraphs explaining who you are and why you're such an interesting person to talk with. Because these lines could be inserted in an introductory letter to anyone you meet on-line, you'd probably like to save them in your UFD.

Then you could go to the SourceMail's Send option. (A handy express-lane command string for that could be entered at the -> prompt: MAIL S followed by the recipient's user ID. That would take you directly to the Subject: prompt.) After you filled in the subject, you could type an introduction, and when you're ready to pick up the prewritten material, you'd type:

****

.LOAD NEWGUY <RETURN>

****

SourceMail would pick up a copy of the text in NEWGUY in your UFD, then prompt you to "Continue." After you finished the letter, you would mail it as usual with the .S command. Voilà—you've just saved yourself more time on-line.

DOTTING YOUR SOURCEMAIL

And speaking of SourceMail, no doubt you've noticed that some of the commands used in SMail are preceded by a period. (Computer folks sometimes call these "dot commands," as in "dot-LOAD" and "dot-S.")

Here's an important point in understanding SourceMail: while you're in the middle of writing a letter, *all* your commands must be preceded by a period. Once you've selected the Send option, SMail assumes everything you're typing is part of the text—unless it comes upon a new line that begins with a period.

So far, you've met two of SourceMail's "dot commands":

.S sends the letter after you're finished with it.

.LOAD, followed by a file name, picks up a copy of a file from your User File Directory.

But there are many other useful dot commands. Let's look at some of them. Again, this isn't something you need to memorize. Just keep the list handy so you can try them out the next time you're exploring SMail on your own. They are:

.DI, which displays the text of your letter. While you're still writing, you can type .DI <RETURN> on any new line to get the full text of what you've written so far.

.DI TO, which displays to whom the letter is to be sent, that is, what user ID or IDs you specified when you filled in the To: prompt.

.TO, followed by a user ID, which adds another recipient to the list, so that he or she also will get a copy of the letter. You can also remove an ID from the To: field by placing a minus in front of it, as in .TO –BBP112.

.DI SU, which displays the Subject: information, and .SU (followed by a new subject line), which changes the subject.

.Q, which quits SourceMail without sending your letter.

.CC, followed by a user ID number, which sends a "carbon copy" to another user.

.EX, which sends the message "express"—that is, it goes to the top of the recipients' mailboxes.

.PA or .PASSWORD (followed by a code word), which gives a message a password. The recipients must know that word to be able to read the message.

.AR, Acknowledgment Requested, which sends you an electronic "receipt" when the recipients pick up the message. If you're sending the letter to a number of people and you want an acknowledgment for only one, you can enter .AR followed by that person's user ID.

Meanwhile, there are some other interesting tricks that can be performed with SourceMail, including tools for mass-distribution of SMails. If you're interested in that, be sure to see the section on Mailing Lists and Distribution Lists in the On-line Survival Kit in the back of the book.

USING THE EDITOR IN SOURCEMAIL

Next to .S, which sends the message, probably the most important dot command in SourceMail is:

****

.ED

****

Entered at the beginning of a fresh line while you're writing your letter, dot-ED summons the EDITOR program.

After you enter .ED, you'll receive the "EDIT" message from the system. Then you can use all the editing commands we looked at while we were on-line—T for TOP, B for Bottom, the Change and Print options, etc.

When you've finished editing the letter, enter FILE <RETURN>. That sends the EDITOR away and puts you right back in the SourceMail program, where the dot commands—including .S to send—take over again.

SUMMING UP

We covered a great deal of ground in this chapter in three major areas—command files, editing, and SourceMail.

You created your first CO file, called SHH, and saw that you can store command strings and then execute them by entering CO and the file name at the –> prompt. We'll start looking closer at these powerful utilities in the next chapter.

You covered the other side of SourceMail—writing and sending—and collected a number of "dot commands" for manipulating the letters you write.

At the heart of the chapter you concentrated on editing on-line and learned these major commands:

T which takes the pointer to the Top of your file.

B which takes the pointer to the Bottom of the file.

n or N# which moves the pointer to the Next line or Next specified number line.

N-# which moves you up # lines. For example, N-2 would move the pointer back up two lines.

L text, which Locates the first reference to a specified word or phrase in your file and moves the pointer to that line.

F word, which Finds the first line that *begins* with a specified word and moves the pointer.

W (Where), which prints the number of the current line, the line being pointed to.

P or P#, which Prints a specified number of lines in your files.

MODE NUMBER, which numbers the lines of the text.

<RETURN>, which toggles from EDIT to INPUT mode.

C/old word/new word, which Changes a word or phrase in the current line to a new entry.

C/old word/, which will remove a word or phrase from the line (that is, Change the old word into nothing).

D, which Deletes the whole line you are on.

I new text, which Inserts a new line of text below the present position.

R new text, which Replaces a line with new text.

/G#, a Global change command, which may be added to the end of a Change command. For example, C/they're/there/G10.

FILE, which files the text, replacing any previous versions.

CHAPTER 7

Finding Folks to Talk With

Now that you have all the tools you need to express yourself, let's find out where you can use them. SourceMail and CHAT aren't the only ways you have to reach out to your fellow travelers. This chapter will begin our search for forums for sharing information.

We turn to The Source for all kinds of communications: buying, selling and trading, spreading the news (and gossip), taking to the soapbox to speak out on an issue, or just to meet new friends and stay in touch with old ones.

In a minute we'll go on-line to see POST, the system's multifunction bulletin board. Some consider it a nerve center of this electronic community. In a way, it's like the classified ad section of your local newspaper, except that it reaches the entire world.

We'll also take a look at some of the newsletters your fellow subscribers produce. The Source offers a unique opportunity for you to become an electronic publisher, to write and share a regular on-line magazine. Before this book is done, you'll learn how you, too, can be a budding William Randolph Hearst.

Then, in Chapters 10 and 11 you'll see how computer communications can rise to a whole new level of sophistication. PARTICIPATE is a state-of-the-art conferencing program. Some of the people you'll meet there are truly making computer network history.

As always, we'll also use these excursions to broaden your knowledge of how to use the system itself more efficiently. For instance, we've got much more to say about the command (CO) files you've just met. On this tour you'll see how to preserve an entire search strategy in a file, going way beyond what can be accomplished in a single command string. And that will be the beginning of customizing The Source for yourself.

A LITTLE BLUE-SKYING

Let's contemplate that for a minute.

Imagine having a personal robot that could help you gather information, reading your mind and *knowing* what you were looking for. We haven't come quite that far, but The Source is taking steps in the right direction.

What if, when you logged on, the system said to itself, "Why, that's ol' Bob. Let's set up his 'usual.' He usually wants to check his stock portfolio. We'll fix it so he can do that by just pressing the 1 key. Also, he'll want to see how his Cubs are doing today—the 2 key will take care of that. And, lately, he's been involved in that hot discussion on PARTICIPATE; number 3 on the menu will let him catch up on what he's missed. . . ."

That particular technology *is* available and many veteran Source subscribers use it daily to get more out of every minute they spend on-line.

All of the systems "tricks" you've been learning are the building blocks we'll use to enhance the way the system deals with you. In the end, it will look as if The Source were designed just to meet your needs. This chapter, then, heralds the setting of a new goal: constructing your own super menu for the features you want to visit regularly. Over the next seven chapters we'll continue collecting the tools. Then in Chapter 15, we'll put it all together.

What you should be doing in the meantime is watching the system in a different way. When you see a new feature—either on our tours or on your own field trips—ask yourself if it's something you'd like to visit regularly. If so, make a note of where the service is and how you got to it. Chances are you'll be able to include it in your customized menu.

CHECKING IN

Log on as usual and, at the menu, go to Command Level.

We want to turn off CHAT, as always, and to do that we can use the new SHH file we saved last time. However, we don't have to execute *all* of the CO file—no need to go all the way to the UPI feature, for instance. This will show you how you can interrupt a command file by sending the BREAK signal.

At the prompt, type CO SHH <RETURN> and the system will begin executing the commands you've stored there. First comes the CHAT -OFF message, then MAILCK. This time, after you get the report of the status of your mailbox, enter CONTROL P. After a moment, the system should report:

****

```
QUIT.
```

****

Notice that we had the system act on the first two steps of the CO file, then bailed out with the CONTROL P before the final step (searching the UPI wires for computer news). Stash that information away; it's another handy use of CONTROL P.

Meanwhile, suppose you've decided that checking the UPI wire isn't something you want to do right after log-on. You now have all the information you need to remove that part of SHH by editing it out. Let's do that. At the prompt, type ED SHH <RETURN>. Then enter P999 <RETURN> (to print 999 lines—or to the end). The system displays the whole file:

****

```
.NULL.
CHAT -OFF
MAILCK
UPI N G COMPUTER
BOTTOM
```

****

The EDITOR's pointer is now at the bottom of the file. Type P <RETURN> and the system will display ".NULL," meaning you are below the last line.

To delete the UPI command string, then, you need to move the line pointer back up a line. Enter N-1 <RETURN> and the system displays the "UPI N G COMPUTER" line. Delete it by typing D <RETURN>.

Now when you return to the Top of the file (the T command) and display it (with P999 <RETURN>), you should see only the CHAT -OFF and MAILCK instructions.

To save this edited version of SHH, type FILE <RETURN>. Remember, that deletes the previous version and saves this latest one. (Had you wanted to leave the EDITOR without saving the changes, you could have entered QUIT.)

One more thing. SHH is now beginning to grow beyond its original purpose of just hushing CHAT. It will grow even more before we're done this evening. So while we have a chance, let's rename it something equally short, but a little more descriptive. How about "GO," suggesting these are the first things we want done when we log on?

At the -> prompt, type

****

CNAME SHH GO <RETURN>

****

meaning, "Change the Name of SHH to GO." CNAME is the command you can use to rename any file in your UFD.

DROPPING BY THE MAGAZINE RACK

We'll now head out for one of the main attractions of this tour, Member Publishing.

As always, we like to give you an idea what part of the system you're traveling to. The best way to do that is to see it first through the menus. So, at the Command Level prompt, type MENU <RETURN> to get back to the top of the system.

****

MENU OF SERVICES <MENU>

1 Today From The Source <TODAY>
2 News, Weather and Sports <NEWS>
3 Business and Investing <BUSINESS>
4 Communication Services <COMM>
5 Personal Computing <PC>
6 Travel Services <TRAVEL>
7 Shopping, Games and Leisure <HOME>
8 Member Information <INFO> **FREE**

Enter item number, <H>elp or <Q>uit:

****

We're going to the Communication Services section of the system again, so type 4, and The Source will display something like this:

****

COMMUNICATION SERVICES <COMM>

1 SourceMail
2 Bulletin Boards
3 Chatting Online
4 Computer Conferencing <PARTI>
5 Member Directory <MEMBERS>
6 Members Currently Online <ONLINE>
7 E-COM and Mailgram Messages
8 Member Publishing <PUBLIC>
9 How to Upload Files to The Source

Enter item number, <H>elp or <Q>uit:

****

Member Publishing is Option 8. Notice the menu is saying you also can reach this area by typing PUBLIC <RETURN> at Command Level.

Type 8 <RETURN> and the system will show you a rather long disclaimer that says all these electronic magazines are the products of subscribers and aren't guaranteed by The Source. After that, there's a menu something like this:

****

WELCOME TO USER PUBLISHING

1 PUBLICATION CATEGORIES
2 MEMBERS-ONLY PUBLICATIONS
3 ALL ABOUT USER-PUBLISHING
4 COMMENTS BY READERS/WRITERS
5 BEST-SELLERS
6 NEW PUBLICATIONS
7 PUBLICATION INDEX

Enter item number or HELP

****

Incidentally, this menu may look a little different by the time you pass this way. Source officials are working on some new displays. Don't let that throw you. Remember that the rules for navigating the system won't change, even if occasionally a feature gets a facelift.

Check your screen for the option that takes you to the Publication Index. It's Option 7 on our screen. Select that option now and it will give you the latest list of all the magazines available. Select it for a list of each publication's name along with its "location number" and the user ID of its publisher.

By the way, a copy of this index is in the On-line Survival Kit at the back of the book. Of course, this, too, is frequently changing, as new publications come on-line and old ones are discontinued. For you, that means several things. You should drop in regularly to look at the "New Publications" feature (Option 6 on our menu). Also, if your software and hardware can accommodate it, it's a good idea to get a printout of this list from time to time to keep up-to-date.

All of the publications being listed on your screen now are accessible from menus. Option 1 would give you "Publication Categories" and all the newsletters would fall into one or more of those categories.

However, as always, the system will save you some time. That's where these location numbers come into play for a special kind of command string. From the -> prompt, you can type PUBLIC followed by a space and the location number of the one you're interested in.

Okay, to see how that's done, we're going to break up into independent study groups now. From the menu on your screen, pick out a publication that looks interesting and take note of its location number.

Now return to Command Level with a Q for QUIT, and at the prompt, type PUBLIC <space> and the location number (such as, PUBLIC 999 <RETURN>. After a few seconds, the system will start displaying:

```
*************************************************************************

    User Publishing programs are produced, maintained and supported
    only by the subscribers whose accounts are listed with each title.
    Like all . . .

*************************************************************************
```

You know where you are now. This is the same disclaimer you saw when we paged through the menus. After the disclaimer, the system says something like:

****

One moment, please . . .

MEMBER PUBLISHING

Version 6.0

(C)opyright Source Telecomputing 1985

Logging BBZ824 . . . .

****

After this message comes the publication's title page. All of the title pages are different, of course, since we've all selected different newsletters. A title page should give you the publication's name, a little information about its specialty, perhaps the name of its editor/publisher.

Following that, you probably see a list of options in a menu form. These could be individual articles, BASIC program listings, tips and suggestions, letters to the editor. The Source gives the publishers a lot of leeway for originality here, so there are many different kinds of menus. Here's a popular style of menu:

****

FILE NAME	LIST TIME	TIMES READ	LAST READING	
1 NEW-HELLO	1:00	25	85–03–17	15:52:17
2 INDEX-HEADLINES	0:54	55	85–03–17	16:22:51
3 THIS WEEK	2:34	121	85–03–17	16:34:41
4 LAST WEEK	2:28	111	85–03–17	16:35:46
5 THE-MAILBAG	5:48	46	85–03–17	16:36:46

Enter a list of options, RETURN for menu, <H>elp or <Q>uit:

****

The publisher indicates you can see any of the articles by typing its number at the prompt. The menu gives you the title or description of the article, an idea of how long it takes to read in minutes and seconds (at 300 baud), the number of times it's been read by other users, and the date and time of the last reading.

In most publications, typing Q <RETURN> at the prompt will take you out of the publication. Tapping the RETURN key (leaving the prompt blank) usually will redisplay the menu.

We'll take a breather here while you explore the publication you've selected. Pick an article or two to look at. As you read, remember you can use your CONTROL codes: CONTROL S to freeze the display; CONTROL Q to restart it. Sending CONTROL P or BREAK usually will interrupt the article and return you to the menu prompt.

We'll be waiting for you back at Command Level. When you're finished, exit the publication and type Q at the next menu you see to return to the -> prompt.

BACK AT THE RANCH

While we're waiting for everyone to get back on the bus, here are a couple of more tips about reading things in the Member Publishing area.

First, there's an even faster way to get to your favorite newsletters. If you visit several of them in one session, you can get mighty tired of that long disclaimer. It will appear at the front door of each one you read. To sidestep it next time, try this

****

-> PUBLIC 123 DIRECT

****

substituting the location number you want for the "123." The "DIRECT" after the location number will cause the system to bypass the disclaimer altogether.

Member Publishing is a real entrepreneurial haven. It costs you nothing extra to read most of the publications. However, the electronic publishers get a piece of the action for the money you spend on connect time when you use their features. So it's to his or her advantage to keep the newsletter as fresh and inviting as possible. You'll find Member Publishing can be a lively regular feature and a rich territory to explore on your own.

How do these folks get paid, anyway? The publishers under contract with The Source receive a monthly royalty based on a percentage of the total billed usage generated by their publications.

If you're looking for what's hot, keep an eye on the Best-Sellers list—see Option 5 in our example back there.

Finally, can you become a publisher? Yes, you sure can. However, The Source suggests you start out by practicing with something it calls "Sharefiles." We'll be looking in on that in Chapter 12. For now, in a nutshell, Sharefiles is a feature that lets you open up a section of your User File Directory for public viewing. You can fill your Sharefiles area with interesting material, spread the word about it, and keep a log of usage to prove there is an interest. Later, if The Source agrees the interest is big enough, you can seek a publication contract and you'll be given some space on the system for your new baby.

But how to do you go about spreading the word?

Glad you asked that.

EXPLORING POST

Our next stop is another feature from the Communication Services menu, from which we seem to have picked most of our targets of late. It's the POST bulletin board. This will give you an idea of what's going on around the system tonight. And we'll have a chance to play again with The Source's searching capabilities.

By now, you certainly know how to get there through the menus. You *could* enter MENU at the -> and follow the Communication Services branch, but we're beyond that now.

At the -> prompt, just type POST <RETURN>. The system will ask, "Do you want instructions?" In the interest of time, type N <RETURN> and you will be shown this prompt:

```
****************************************************************

<R>ead, <PO>st, <PU>rge, <S>can, or <H>elp:

****************************************************************
```

It looks a little like SourceMail's prompt and the one in the Member Directory, doesn't it? And it works the same way—you're being asked for one or two letters to indicate the option you want.

The service lets you read or scan messages already posted by others and post or purge your own messages. Let's scan first. Type S <RETURN> and the system displays:

****

<C>ategory, <U>ser ID, <D>ate, <K>eyword:

****

Like the Member Directory program, POST lets you home in on the messages of interest. The last three options—user ID, date, and keyword—are familiar, but what's this "category" stuff?

Let's try it and see. Type C <RETURN> at the prompt and see:

****

Categories, or <H>elp:

****

The system is waiting for a word or two.

Messages on POST, in addition to having keywords, are assigned to categories by the people who have posted them, just as the ads in a newspaper classified section are listed under headings such as "Help Wanted" and "Apartment for Rent." Searching by categories can speed up your on-line browsing considerably. If you were interested only in messages about Kaypro computers, for instance, you could specify that category.

Typing H <RETURN> here would provide a list of the latest categories. If you'd like to do that, help yourself, but it may take a minute or two, so allow us. These are the categories we retrieved the last time we were here:

****

Antiques	Apartments
Apple	Atari
Automobiles	Aviation
Books/publications	Bulletin-board
Businesses	CADO
Chatter	Clubs
Collectables	Commodore
CP/M	CPA
CTSLATE	DEC
Dimension	Documentation

Education
Epson
Games
Gripes
Ham-radio
Hardware
Hayes
Heath
Help-wanted
Hewlett-Packard
Hobbies-and-crafts
IBM
Investments
Kaypro
Macintosh
Merchandise
Music
NEC
Novation
Office-equipment
Osborne
PARTI
Personal
Photography
PLATO
Politics
Programming
Public
Public-files
Real-estate
Services
Soap-operas
Software
Source
Sports
Tandy
Taxes
Teens
TI
Transend
Travel
Victor
Video
VisiCalc
Xerox
Zenith

****

Since you're now used to changes around the system, it shouldn't surprise you that the list of categories is constantly being added to and sometimes categories are removed for lack of interest. Therefore, from time to time you should get a current list from the Help option. We'll even show you an express-lane command for that in a few minutes.

For this test, let's look for notices in the "Apple" category. At the prompt on your screen, type APPLE <RETURN> and the system, after a second, will reply:

****

Searching. . .
150 notices valid.
<N>arrow, <E>xpand or Return for all :

****

Obviously, you'll find a different number of Apple notices in your search.

The message has ended with a familiar prompt. You last saw it in the Member Directory and it works the same way here. You can either Narrow the group (reducing it from 150) or Expand it by defining another category to include.

We'll whittle it down a little. Type N for Narrow.

```
****************************************************************
     <C>ategory, <U>ser ID, <D>ate, <K>eyword:
****************************************************************
```

This time, we'll take the Date option—type D <RETURN> for this prompt:

```
****************************************************************
     Starting and ending date or <H>elp:
****************************************************************
```

A few notes here:

1. Messages on POST remain public for seven days, just like the stories on the UPI wires.

2. You can enter a specific day and month, such as 16 MAR, or you can request all the entries for that month with just MAR (assuming, of course, that the current month *is* March).

3. Tapping RETURN and leaving the prompt blank will give you just today's messages.

4. If you enter only one date (within the past seven days), then all notices between today and that date will be selected.

Okay, press RETURN for today's Apple notices. When we did it, the system reported:

****

Searching. . .
17 notices valid.
<N>arrow, <E>xpand or Return for all :

****

Now tap RETURN again to see what the scan looks like. You should be seeing something like this:

****

Wait. .
1 18 MAR TCV000 MAPPING PROGRAM?(APPLE)
2 18 MAR AAD000 PROGRAMMER AVAILABLE(APPLE)
3 18 MAR TCY000 APPLE PROGRAMS FOR SALE(APPLE)

****

The list eventually will end with:

****

Enter item(s), <H>elp, or Return for all :

****

To see what the messages themselves look like, pick a number from the menu on your screen. You could also pick more than one message, typing in all the message numbers and separating them by commas, such as 1,5,9. For each, you'll be shown the message "header," something like this:

```
**************************************************************

Category:APPLE
Subject:APPLE PROGRAMS FOR SALE
From:TCY000
Posted:18 Mar 8:27 pm

<N>ext, <PO>st, or Return for text- :

**************************************************************
```

The prompt gives you three options: N will take you to the next message; PO will allow you to post a message of your own at this point; <RETURN> (leaving the line blank) will display the text of this message.

Not shown here are the two commands that you've discovered will work just about anywhere in The Source: Q to QUIT and return to the -> prompt, and P to go to the Previous menu. Even though they're not listed, they will work here.

Do this—look at the text of the message by tapping the RETURN key. When you come to a -More- or an -End- prompt, type P for Previous and the system reminds you that there are:

```
**************************************************************

17 notices valid.

<N>arrow, <E>xpand or Return for all :

**************************************************************
```

Enter P again and up comes:

```
**************************************************************

<C>ategory, <U>ser ID, <D>ate, <K>eyword :

**************************************************************
```

All we're doing here is just backing up through the levels of options that we took in the first place. If we continued typing P's, we'd eventually get right back to the top of the POST program.

But instead, let's retire to Command Level and approach this feature from a different direction. Enter the Q command.

REGROUPING

Before we move on, let's go over a few facts that will make it easier for you to explore POST.

First, the procedure for reading messages is essentially the same as for scanning them. You could just select the Read option instead of Scan. The only difference is that the Read option takes you directly to the message headers as soon as you've narrowed or expanded your search. Scan is handier if you expect to find a large number of messages, since it will build a menu.

You saw an opportunity to enter keywords in your search. These differ from categories—they don't have to be on a prefiled list in the system. Keyword searches are particularly useful because they will call up all notices that contain the words you designate in *either* their category or subject lines. In other words, if you searched for the keyword rather than the category of "Apple," you'd find all the messages in the Apple category as well as those in other categories that mention "Apple" as part of the subject line.

And you can get additional on-line information about POST features by typing HELP and a command at any prompt (such as, HELP POST or HELP READ or HELP PURGE).

ABOUT POSTING AND PURGING

To leave your own notices on POST, simply select the PO option from the main prompt:

```
**********************************************************************

&lt;R&gt;ead, &lt;PO&gt;st, &lt;PU&gt;rge, &lt;S&gt;can, or &lt;H&gt;elp:

**********************************************************************
```

After that, the system will ask you for the category under which you want the notice published. Refer to that same list we showed you earlier or type H <RETURN> at the prompt for the latest list. Try to select the category that best suits your subject; your fellow subscribers will be using that option to home in on *their* interests.

After that, you'll be asked for a brief subject line—you have only 32 characters, so use simple and direct words, like PUBLIC PROGRAMS FOR SALE. As you've seen, these words will be displayed in the header when others read your message.

Finally, write your message. Don't exceed 78 characters on a line. You have up to 23 lines to have your say.

When you're finished, type

****

.S <RETURN>

****

and the system will take it from there. New messages usually are posted within a few minutes. By the way, you can post identical messages in different categories, but The Source asks that you limit yourself—identical notices should be in no more than three categories at one time.

As with SourceMail, you can use the EDITOR program when writing messages on POST. Before entering the dot-S command, you can type .ED <RETURN> on a new line. That will take you to EDIT mode, where most of the editing commands you've already learned will work. That includes the dot-LOAD command, so you can post a file that exists in your User File Directory. Just remember that the total message can be no longer than 23 lines of 78 characters. When you're finished editing, press the RETURN key on a new line (exiting EDIT mode), then dot-S when you're ready to post it.

Another option at the main prompt is PU for PUrge. This will let you delete any messages you've previously posted. When you enter PU at the prompt, the system automatically will scan the bulletin board and display a menu of the messages it has under your user ID. Then you can delete them by number.

POST IN THE FAST LANE

You've come to expect that The Source will always have some shortcuts to its features and POST won't disappoint you. Everything we've done so far, and much more, can be performed through command strings and CO files.

Slip a bookmark on this page—we'll be referring to it again—and then try this. At the -> prompt, type:

****

-> POST S K APPLE <RETURN>

****

After a moment, the system should report:

****

Searching. . .
150 notices valid.
<N>arrow, <E>xpand or Return for all :

****

This is in the same family as the other command strings we've met—UPI N G COMPUTER to search the news wire for computer-related stories, MAIL R UN to read previously unread SourceMail, etc.

Now, if you're still shaky about the logic we're using here, promise yourself you'll do a little homework. After we're off-line, back up a few pages and compare this command string with how we used POST the first time around. What this string is really doing is (1) summoning the POST program, (2) choosing S for <S>can at the first prompt, (3) choosing the <K>eyword option at the next prompt, and finally, (4) defining "Apple" as the keyword.

Three notes:

1. We chose to Scan a Keyword rather than the Category because, as we just found out, the Keyword option searches both the message category and subject lines for the phrase we've specified.

2. You could just as easily Read the messages as Scan them, if you modified the command string to POST R K APPLE.

3. If you wanted to Scan or Read by Category, instead of by Keyword, you could enter the command string as POST S C APPLE or POST R C APPLE. However, the system also will let you shorten that if you'd like. If you're scanning by the Category of Apple, you need enter only POST S APPLE. In other words, The Source will "assume" you mean the Category unless you specify "Keyword" with the K.

The next prompt is the one that lets you Narrow or Expand the group. Type N <RETURN> and at this prompt

****

<C>ategory, <U>ser ID, <D>ate, <K>eyword:

****

select the Date alternative by typing D <RETURN>. When The Source says

****

Starting and ending date or <H>elp:

****

just press the RETURN key for today's notices.

STORING THIS STRATEGY

From here on, everything is predictable. As we found last time, the system will pare down the selected group to today's notices and give you an option to expand it, narrow it further, or make a menu of what you've found so far. Therefore, we don't need to walk all the way through that again.

Instead, come on back to the Command Level by typing Q at the next prompt and let's compare notes.

This kind of search of POST—that is, picking up the latest notices about a subject of interest—is a very popular one among Source regulars. And because the command string is always the same, the string is an ideal candidate for storing in a CO file for easy access. Perhaps you could call it POSTCK. At the -> prompt, open a file by that name:

****

-> ENTER POSTCK <RETURN>

****

After you've received the "Enter text" message, you can begin by typing the first command string. (This is where you might want to refer to the bookmarked passage. This file ultimately will reenact that whole procedure.)

Your first line is:

****

POST S K APPLE <RETURN>

****

(If you want to Scan for something other than APPLE, put in another word. You can even have two or three keywords, separated by commas, such as POST S K APPLE,TANDY,IBM. The more keywords you have, the more messages you'll find in the search.)

As you saw when we walked through this demonstration, you're next asked to Narrow or Expand the group. We want POSTCK to handle this for us, and to Narrow the bunch down to today's offering. So on the next line, type:

****

N <RETURN>

****

The next prompt POSTCK needs to anticipate is the one asking how we want to narrow the field—<C>ategory,<U>serID,<D>ate, <K>eyword. On the next line, then, type

****

D <RETURN>

****

for date.

Here's a neat trick. As you see if you're looking at the section we asked you to bookmark, the next prompt we'll come to is looking for the date specifications. On the earlier run-through, we just pressed RETURN to signal "today." The CO file can do that, too, if you leave a space on a line all by itself. So on the next line of POSTCK, enter one space followed by a RETURN, like this:

****

<space> <RETURN>

****

We'll do that again on the next line. That's because at that point, The Source will have narrowed our group down to today's notices and will be asking if we want them displayed. Yes, so on the next line, type another:

****

<space> <RETURN>

****

Now close POSTCK by pressing the RETURN key on a new line and you'll be back at the -> prompt.

If you wanted to admire your handiwork, you could type TY POSTCK <RETURN> and the system would display

****

```
POST S K APPLE
N
D


```

****

with two blank lines trailing after the D.

Now to see it do its stuff, just enter CO POSTCK <RETURN> and stand by.

Nice how one little five-line CO file can have the system standing on its head for you, isn't it? You should be brought to the top of the list of latest messages on your chosen topic (APPLE or whatever) with the system ready to display them for you. Read a few, if you'd like. Then meet us back at the Command Level prompt when you're done. We're not finished with POSTCK.

PULLING IT ALL TOGETHER

Now since this searching of POST is so handy, it would make sense if you wanted to make it one of the first things you did each time you logged on—that is, make it part of your GO file.

Easy enough.

At the -> prompt, call up GO for editing by typing:

****

-> ED GO <RETURN>

****

After you receive the "EDIT" message, enter a B to go directly to the Bottom of the file. Then type:

****

LOAD POSTCK <RETURN>

****

In a second, the system will display another "EDIT" message. Let's take a look at the newly merged file. Enter T to go to the Top, then P999 <RETURN> and you should see:

****

```
.NULL.
CHAT -OFF
MAILCK
POST S K APPLE
N
D


BOTTOM
```

****

Getting more sophisticated all the time. (The Load command hasn't deleted the original POSTCK—it just moved a copy of it to GO.)

Now type FILE <RETURN> to save the new version of GO. In a minute, we'll see it run all over the system performing tasks for us. But first, let's have a little fun with it.

COMO (WHAT?) COMO

We have time for one more new command before we call it quits for the night. This one, called a Command Output, or COMO, file is another one for your bag of systems tricks and works like an on-line staff photographer.

Here's the background. Anything that can be viewed on your screen also can be captured in a file and stored in your User File Directory, just as if you had written it there yourself. That means files you copy with COMO also can be edited and reused on the system. Many Source subscribers who have an active correspondence through SourceMail, for instance, use COMO to capture and share interesting material they've found on POST and other areas of the system.

The only requirements are that you (1) turn on COMO before you go to the area you want photographed, and (2) return to the Command Level afterward and turn COMO off again.

Now, here's how it works:

Suppose you wanted a copy of any new POST notices about IBM equipment. You could start out by opening a COMO file. At the Command Level prompt, you'd type COMO IBM <RETURN>. (But don't do that now—we'll practice with COMO in a minute.)

After the IBM file has been opened this way, *everything* that appears on your screen, either sent by The Source or by you from your keyboard, will go into the file named IBM.

Then you'd just enter a POST command as usual, such as POST S K IBM <RETURN> and follow the prompts as you have before. When you had everything you wanted in the file, you'd return to Command Level with the Q command and type COMO –E <RETURN>. Note the space between COMO and the hyphen. That would tell The Source, "That's the end of the COMO session." The system would respond with another –> prompt.

COMO will work anywhere in the system to collect material for you. You can even use it in connection with a CO file.

At the prompt on your screen now, type

****

–> COMO NEWSTUFF <RETURN>

****

(meaning, "Open a file called NEWSTUFF"). Now just type CO GO <RETURN> to activate your GO file. It turns off CHAT, runs the MAILCK, then goes to POST and scans for all notices about APPLE, narrows the list down to those posted today, and finally starts showing you that list. (Incidentally, if there were no new notices about APPLE, the GO file would give you all previous messages with APPLE as a keyword.)

GO stops when it runs out of commands. Read a few more messages if you'd like. When you're done, enter the Q command.

Now—very important—turn off the COMO file. (If you don't do this, you'll continue collecting material in NEWSTUFF wherever you go in the system.)

To turn it off, type:

****

–> COMO –E <RETURN>

****

That's COMO<space> –E.

And now you should have a faithful record of your GO excursion preserved in the NEWSTUFF file. You could look at it with TY NEWSTUFF.

When you've seen enough—and you probably are getting pretty tired of this same stretch of road by now—type a CONTROL P to BREAK and you'll come back to the Command Level prompt.

Incidentally, some users like to "time stamp" the information they collect in a COMO file. Remember the DATE command we learned on the first tour? You could make that the first command you enter after COMO is initiated. It'll keep a record of when the information was retrieved.

WASHING UP

Before you log off tonight, you'll need to delete the files we created so you won't be charged for their storage. All we really want to keep is GO, so type:

****

-> DEL NEWSTUFF,POSTCK <RETURN>

****

That concludes the scheduled part of our tour. However, before you log off, you might want to continue your independent studies by exploring POST a little further.

POST is the best feature we've come to so far for practicing with the powerful command strings. And the system even has a little shortcut if you want to browse by a specific category. At the -> prompt, you can type POST followed by either S or R (to Scan or Read) and one of the categories we mentioned earlier. For instance, POST S (or R) MUSIC will find notices about music, and POST S (or R) SOAP-OPERAS will reveal news of the world of daytime TV.

If you want to put in a keyword rather than a category, you should precede it with a K, such as POST S K TEACHER, etc.

If you feel up to a little adventuring, then, refer to our category list earlier in this chapter and go to it. You might even want to try making new CO and COMO files. You have enough information to experiment. Remember to delete them before you log off.

Have fun, and when you're finished, we have some closing thoughts.

ODDS AND ENDS

By the way, we've been telling you to use the TY command to "type" or display a file in your UFD. There's another command for this—CRTLST. It works just like TY except that the display stops every twenty-four lines and shows you a -More- message. If you had a long file called, say, LONG.FILE, in your UFD, you might type

****

-> CRTLST LONG.FILE

****

to have those helpful pauses available.

On the other hand, sometimes those -More- messages can be a nuisance when you run across them in an article you're reading. Some Source subscribers prefer to turn them off and depend on CONTROL S and CONTROL Q to stop and start. Often you can overrun the -More- prompt. Try this sometime: type NOMORE <RETURN> at the first -More- you see in an article and you shouldn't see another -More- until you reach the -End-.

WRAPPING UP

We saw two important areas of the system on this tour—Member Publishing and POST—both of which deserve some further exploration on your own.

POST, in particular, will be a feature you'll return to, we think. It's a touchstone for many subscribers who want to spread the word about interesting events going on elsewhere on the system. For instance, you probably noticed that one of the categories we listed for POST was PARTI. When you move on to the PARTICIPATE feature in Chapters 10 and 11, you'll see how PARTI enthusiasts use that category for important announcements.

And speaking of special categories, there are a couple of others you should know about. One is SOURCE; another is GRIPES. These categories are regularly monitored by The Source's customer-service people. They give you a public place to discuss ideas and problems with Source officials. Often suggestions for new features on the system are brought up in the SOURCE category. And it may happen that questions nagging you about the system will be answered on the GRIPES board.

The electronic entrepreneurs over in Member Publishing also advertise on POST. There is a special category for them: PUBLIC-FILES. We usually try once a week to scan POST with a keyword of PUBLIC, as in

****

-> POST S K PUBLIC <RETURN>

****

to keep tabs on new publications. Also there you'll find notices from subscribers who have opened up their Sharefiles areas for test publications. (More about using Sharefiles in Chapter 12.)

Also, you may have noticed the category called CHATTER. That's a good place to find folks who like to use the real-time CHAT feature. And, of course, if you're feeling talkative sometime, leave a message there, telling the time and date you'll be on-line. Just remember to purge the note after the offer expires.

Earlier we mentioned that there was an express-lane command for getting the latest list of POST categories. Now that you've explored command strings in POST, there's nothing too mysterious about this one:

****

-> POST S C H <RETURN>

****

(In other words, "Call POST, choose <S>can by <C>ategory, and then take the <H>elp option.")

In this chapter we also learned about COMO (Command Output) files. Entering COMO and a file name at the -> prompt will put all subsequent material into a file in your UFD. To close the file, enter COMO -E.

We learned that PUBLIC followed by a location number will take you to a specified Member Publishing newsletter, but that you can get there faster by using the DIRECT command, as in PUBLIC 999 DIRECT. That bypasses the long disclaimer.

And you found a new file-maintenance command, CNAME, for Change Name. It will rename a file in your User File Directory. We used it as CNAME SHH GO to change the name of the SHH file to GO.

CHAPTER 8

Taking The Source on the Road

Enough of all this serious stuff. You've had seven chapters' worth of hard work. You deserve a break today (and probably tomorrow, too). In this chapter and the next, we'll let The Source carry most of the load.

Up until now, we've focused on features in the system's Communication Services branch, learning how to put your ideas into electronic words. We'll be coming back to this important area. For example, later chapters will look at how to use that very powerful computer conferencing program called PARTICIPATE. And before we're done we'll take a peek at the system's newest form of communications, the special-interest groups.

However, we mustn't give you the impression that the system is good only for communicating with other Source subscribers. On the contrary, The Source is much like your own community—there's more to it than mailboxes and telephone poles. Thousands of computer owners have come to count on The Source more for research and recreation than as a means of self-expression.

So, this time we'll show you how to use the system in some new ways, such as how to

- Find good hotels and motels.
- Pick quality restaurants in the towns you visit.
- Locate the best flight for the journey.
- Check out the weather situation so you'll know what to expect.

• Avoid traveling altogether by letting the system help you shop right from the keyboard.

Then, looking ahead to Chapter 9, we'll round out our much-deserved respite by heading to another area and watching The Source at play—the on-line games arcade.

You know, features like these would have seemed like science fiction a few years ago, but today they're just business-as-usual on The Source. Obviously, not all of them will be of interest to you, so feel free to browse this chapter. Sometimes it's enough just to know a service is available if you need it.

For now, settle back—we won't be firing up the computer tonight—and let us introduce you to your on-line service staff.

THE SOURCE AS YOUR SERVANT

Planning a trip and shopping may be two of the most time-consuming things you ever do. While they can be fun, they also can be frustrating if you're under the gun to get the job done in a hurry. The Source can take much of the drudgery out of both chores if you learn how to search its specialized databases.

Both travel and shopping have their own menus in the system. However, as we've said before, be prepared for minor changes in the way menus look. That's especially true here. On-line traveling and shopping services are two of the biggest growth areas, so you can expect to see new features coming on-line all the time.

The Source's Travel section is Option 6 on the system's main menu. Or you can reach it by typing TRAVEL <RETURN> at the –> prompt. Either way, your next menu will look something like this:

```
**************************************************************

TRAVEL SERVICES <TRAVEL>

1 Air Schedules and Fares
2 Hotel and Restaurant Guides
3 Agency Services
4 Accu-Weather <WEATHER>
5 Travel Bulletin Board

Enter item number, <H>elp or <Q>uit:

**************************************************************
```

There are multiple listings under most of these selections and some interact with features you've already seen. For instance, Option 5 here would take you to the POST bulletin board and automatically search for messages in the Travel category, then Scan them and give you a list. That's set up for folks who haven't learned as much about the system as you have—they don't know the express-lane commands. Since you've just completed a tour of duty on POST, you know all about entering commands like POST S TRAVEL at the -> prompt, which will get the same information. Feeling smug?

Now, to see how some of the other features work, let's imagine a scenario and use two of the databases in the Hotel and Restaurant Guides area to help solve a problem.

THE CASE OF THE DUCK-HUNGRY BOSS

You're sitting in your office one quiet afternoon when the phone rings. It's The Boss. He summons you to his office, and from the tone of his voice, it sounds as if something important is about to happen.

You find him sitting behind his desk with that grin that says, "If I were you, I'd just resign now and avoid this heartache."

He explains the problem as you take notes. You are to accompany him to Huntington, West Virginia, for a very important business meeting—day after tomorrow.

You are to make the arrangements for the trip and The Boss has a few requirements:

1. He wants a hotel with both a pool and tennis courts, since he wants to pretend he's athletic and he doesn't know which sport he'll be up for just yet.
2. He tells you his religion requires him to eat duck and to not handle money on Thursday nights, so you'll need to find a restaurant that serves duck and accepts MasterCard.
3. He has no patience—you have until 4:00 P.M. to show him the itinerary.

You wander from the office, your head swimming with strange thoughts. Let's see. Did he say he wanted to swim with a duck and that it had something to do with a religious experience?

Well, with The Source, you can cater to the whims of The Boss and settle the entire matter in fifteen minutes. He may not be impressed—he always says he expects as much from you—but, just between us, he'll be surprised.

FINDING HOTELS AND MOTELS

The menu offering guides to hotels and restaurants looks something like this:

****

1 A-Z International Hotel Guide <AZ>
2 Mobil Hotel Guide <USROOM>
3 Mobil Restaurant Guide <USREST>

Enter item number, <H>elp or <Q>uit:

****

All three of these services are databases that are searchable by keywords. So, as you've done with POST, the Member Directory, and other features earlier, you can use The Source's excellent search utilities to zero in on a hotel or restaurant that meets your specifications.

For this demonstration, we'll use the Mobil Hotel Guide and the Mobil Restaurant Guide. Notice that you can use express-lane commands to get to them by typing USROOM and USREST at the -> prompt.

If we were on-line now and you selected the hotel guide, here's what you'd see:

****

Welcome to the Mobil Travel Guide, your window into fine lodging.

Enter "help" for further instructions.

City,State (for example, CHICAGO,IL):

****

As the prompt says, the first thing the database needs to know is the name of the city and state to which you're going. As always, you may use capital or lowercase letters, but the city and state must be separated by a comma with no spaces after the comma. The state should be entered as the two-letter postal code.

The Boss said the city in West Virginia is Huntington, so you type HUNTINGTON,WV <RETURN> following the colon prompt and, after a moment, see:

****

4 accommodation(s) found in Huntington, WV. 4 total.
City,State or <cr> to search:

****

The program is offering you a chance to select another city and state or search the accommodations you've already found. You chose Huntington, WV (well, *we* chose it, because it's where we live), so now you need to search for a hotel or motel that will satisfy The Boss's first requirement. You press RETURN to see:

****

Select <S>can, <F>ind, <B>ackup, <P>rint,
<R>estart or <H>elp for more instructions:

****

That's similar, but not exactly the same as some of the prompts we've seen earlier. It works the same way, though. The letters in brackets represent commands you can enter. An S shows you the names of the four hotels or motels; an F lets you search for keywords in the descriptions of those four; a B takes you back to the previous command prompt; a P prints out on the screen the complete information about all four you've found; an R lets you start from the top of the program again; and an H gives you more instructions.

The problem at hand is finding hotels and motels with a pool, so you enter F for Find and are shown a prompt that says, "Search for:". You type POOL <RETURN> and here's what you see:

****

All the current lodging(s) have that

4 accommodation(s) with "POOL".

Select <S>can, <F>ind, <B>ackup, <P>rint,
<R>estart or (H)elp for more instructions:

****

Whew. That's one hurdle crossed. Because all of the selected motels or hotels have pools, the program is holding all four in its memory. If only two of the four had pools, the program would have dropped the two without pools and left you with two hotels to work with.

Now you must find which, if any, has a tennis court. You type another F and tell the program to search for "tennis."

****

1 accommodation(s) with "POOL", "TENNIS".

Select <S>can, <F>ind, <B>ackup, <P>rint,
<R>estart or <H>elp for more instructions:

****

The program found only one hotel or motel that offered *both* a pool and tennis courts. For your search, that's the only hotel you really need information about. So you type P <RETURN> (for Print) to see the specifics:

****

The following 1 accommodation(s) have "POOL", "TENNIS" in their descriptions.

Accommodation(s) in Huntington, WV

*** HOLIDAY INN GATEWAY. (Motels & Motor Hotels) (Box 2528; 25725) 6007 US 60E, 8 mi E, 2 ½ mi E of I-64 exits 15, 20. 736–8974. 224 A/C rms, 1–2 story. S $41–$48; D $47–$54; each addl $6; suites $102–$192; kit. cottages $48–$55; studio rms $45; under 19 free; sr citizen, wkend package rates. Crib free, cot $6. TV; in-rm movies. Indoor pool. Playground. Cafe 6 am–10 pm. Supper club 4–11 pm. Rm serv 7 am–11 pm; Sun to 10 pm. Private club 4 pm–11 pm; dancing exc Sun. Ck-out noon. Meeting rms. Gift shop. Barber, beauty shop. Lighted tennis. Health club. Masseur. In-house movies. Bellhops. Some extra basins; many oversize beds. Cr cds: A, D, MC, V. Wheelchairs. Tennis. Swimming.

1 accommodation(s) with "POOL", "TENNIS".

Select <S>can, <F>ind, <B>ackup, <P>rint,
<R>estart or <H>elp for more instructions:

****

Well, so much for that. Quite a lot of information in a very short time.

FINDING RESTAURANTS

Ah, but now you're thinking about the old man's odd Thursday night religious custom, and a familiar sinking feeling comes back. Visions of ducks and MasterCards.

That's where the companion program, the Mobil Restaurant Guide, comes into play. You leave the hotel guide by typing Q. As always, that takes you to the –> prompt, where you can type USREST <RETURN> to get to the restaurant guide.

The restaurant guide is searchable in exactly the same way. After entering "HUNTINGTON,WV" at the first prompt, the program searches for restaurants in Huntington and offers:

****

3 restaurant(s) found in Huntington,WV. 3 total.
City,State or <cr> to search:

****

(Uh, the president of our local chamber of commerce wants you to know we have more than three restaurants in our fair city, but only three are recommended by Mobil.)

Let's see how many of them accept MasterCard. The abbreviation most commonly used for MasterCard is MC. Will that work?

****

Select <S>can, <F>ind, <B>ackup, <P>rint,
<R>estart or <H>elp for more instructions:F
Searching. . .
Search for:MC

2 restaurant(s) with "MC".

Select <S>can, <F>ind, <B>ackup, <P>rint,
<R>estart or <H>elp for more instructions:

****

Very good. Two of the three accept MasterCard.

But now for the acid . . . er, duck test. Entering another F, we perform the famed "duck search" and:

****

Search for:DUCK

1 restaurant(s) with "MC", "DUCK".

Select <S>can, <F>ind, <B>ackup, <P>rint,
<R>estart or <H>elp for more instructions:

****

We've narrowed the search to one restaurant. You type P and the system displays:

****

The following 1 restaurant(s) have "MC", "DUCK"
in their descriptions.

Restaurant(s) in Huntington, WV

*** PERMON'S AT THE TOP. West Virginia Building, 4th Ave at the Plaza, downtown; I-64 5th St exit. 525–8600. Hrs: lunch, Mon–Fri 11:30 am–2:30 pm; dinner, Mon–Thurs 6–10 pm; Fri, Sat to 11 pm. Closed Sun; Memorial Day, July 4, Dec 25. Res recommended.

A/C. Continental menu. Private club 5 pm–midnight; Sat to 1 am; closed Sun. Wine list. Table d'hote: lunch $3.95–$6.95, dinner $12.95–$29.95. Child's plates. Specialties: duckling a l'orange, tenderloin au poivre du chef, tournedos of beef Huntington. Own baking. Background music; pianist; dancing at dinner. Valet parking. On 15th floor; spectacular view of countryside. Jacket suggested. Cr cds: A, MC, V. Wheelchairs. Spoken: German,Spanish,French.

****

By the way, notice this: in the description, the menu item is listed as "duckling a l'orange." In this particular keyword search, the program looked for any word in the description that contains the four letters "d-u-c-k," even if they're in the middle of the word.

However, you'll find that not *all* keyword searches on The Source work the same way. Some search for words, others search for letters within words, and others search for phrases. We'll be devoting an entire chapter later to an in-depth look at searching on The Source, including how the different features use keywords. We think that the more you use these features, the more you'll appreciate the work The Source has done with its searching utilities, so it's worth our time to learn them well.

Meanwhile, imagine The Boss's face when you come back in fifteen minutes with the entire itinerary for his approval. Unless, of course, you'd rather stay in your office for a couple of hours more and let him think you worked your fingers to the bone. . . .

A–Z ONLINE TRAVEL SERVICE

The A–Z Travel Service, on the same menu with the Mobil features, is another—but larger—searchable database. It contains up-to-date listings of more than 23,000 hotels worldwide. Information on each hotel includes the street address, the location (such as downtown, airport, etc.), the local telephone number, a toll-free number, if one is available, and ratings. Other information includes the telex number, the number of rooms available, the regular rates (in U.S. dollars), credit cards accepted, facilities and business services available (such as meeting and convention rooms).

With it, you have a little more flexibility in the use of keywords. For instance, to save you time, A–Z lets you string them together, separated by commas. That means you could type POOL,CONF,TV <RETURN> at the keyword prompt.

Because this is such an extensive database, you may find it a little more difficult to use than the Mobil Hotel Guide, but not much. And if you intend to use it often enough to become familiar with the commands, and particularly if you're looking for hotels overseas, the A–Z Hotel Guide will be worth a look. Incidentally, on-line help is available there by typing I for Instructions immediately after you enter the program.

THE WEATHER ON THE SOURCE

No matter how you travel, knowing the weather at your destination is important. And if you're traveling by car, you also may need to know the weather between "here and there." That's why The Source's Accu-Weather reports are included on the Travel menu.

Here's how it works. When you choose the weather option or type WEATHER at the command prompt, you'll see something like this:

```
*****************************************************************

ACCU-WEATHER (TM)
Copyright Accu-Weather (TM) 1984

MAR 03, 1985

1 NATIONAL FORECASTS
2 REGIONAL FORECASTS
3 MAJOR METRO AREA FORECASTS
4 CITY FORECASTS BY STATE
5 WORLD WEATHER
6 ACCU-WEATHER FEATURES
7 REGIONAL HIGHWAY FORECASTS (NOV–MAR)

<I>nstructions

Enter item number or <H>elp:

Enter State, two letter code or <A>ll:

*****************************************************************
```

This feature also uses those two-letter postal codes for the state abbreviations. Type a code after the colon prompt, and the system will give you the forecasts it has on file for cities in that state.

For example, if you asked for California—by typing CA <RETURN>—you would see:

****

ACCU-WEATHER CITY FORECASTS BY STATE

	THU	FRI	SAT
CA Los Angeles	74/55pc	74/55pc	74/55pc
CA Sacramento	72/45pc	72/45pc	67/40c
CA San Diego	70/55pc	70/55pc	70/55pc
CA San Francisco	65/47pc	65/47pc	60/43c
CA San Jose	70/45pc	70/45pc	65/45c

s-sunny pc-partly cloudy c-cloudy sh-showers r-rain
t-thunderstorms sf-snow flurries sn-snow i-ice

****

That's pretty straightforward.

Perhaps an even more important service of Accu-Weather is on Option 7 of the weather menu: the U.S. regional highway forecasts, which are available in the winter months. Choosing it would show you this menu:

****

MAJOR HIGHWAY ROUTE FORECASTS
(Nov 1–Mar 31)

1 WESTERN REGION
2 CENTRAL REGION
3 EASTERN REGION
4 KEY TO STATES BY REGION

Enter item number or <H>elp:

****

If you'd chosen Option 3 (Eastern Region) on March 3, 1985, you might have seen:

****

03-MAR-85 18:00

INTERSTATE ACCU-WEATHER FOR THE EAST ZONE

I-90 WESTERN NEW YORK
HWY 17 EXTREME WEST NEW YORK
I-81 NORTH NEW YORK
FLURRIES AND BLOWING SNOW WILL END EARLY TONIGHT. DRY LATE TONIGHT AND THURSDAY.

I-87 NORTH NEW YORK
I-89, I-91, I-93 NORTHERN VERMONT AND NEW HAMPSHIRE
I-95 NORTHERN MAINE
GUSTY WINDS EARLY TONIGHT ALONG WITH A PERIOD OF SNOW THAT CAN BRING A COATING TO AN INCH OR SO OF SNOW TO SOME AREAS. WATCH FOR SLIPPERY SPOTS.

ELSEWHERE IN THE EAST ZONE, NO IMPORTANT SNOW OR ICE PROBLEMS EXPECTED FOR AT LEAST THE NEXT 24 HOURS.

****

This service is popular with most of the traveling business people on The Source, particularly now that modern briefcase computers are so affordable. The portables make it easy to log on to the system from virtually any hotel room, so the travelers use this information to stay on top of road conditions on the next leg of the trip.

Of course, these and The Source's other weather forecasts are updated regularly throughout the day.

PLANNING A FLIGHT ON-LINE

Planning an airline flight can be traumatic. Talking to someone on the phone about the confusing web of flights these days is not the greatest way to launch your voyage.

Help is available through two Source services, both accessible from the Travel menu we saw at the beginning of this chapter:

1. Dittler Air Schedules will list domestic and international flights between any two cities you specify and show you other pertinent flight information.

2. The database of the Official Airline Guide (OAG), Electronic Edition, is available on several services and contains more than a million flight schedules for 650 national and international flights. Fare information is available for direct flights and connections for North America, including Canada, Mexico, and the Caribbean.

OAG is a value-added service, which means there's a surcharge beyond the normal Source costs. However, you can read about OAG and look at its background information on-line without incurring the surcharges.

Each of these services works a little differently but we can illustrate the basic method for using either.

Suppose we were to search the Dittler Air Schedules for flights between Cincinnati, Ohio, and Orlando, Florida. Choosing the domestic flights listing from the appropriate menu (in this case, it's called AIR-SCHED-D, with D meaning "domestic flights"), you'd see something like this:

****

Domestic Air Schedules (USA, CAN, MEX,)

-C- Copyright 1982 Dittler Brothers Inc., Atlanta, Ga. All Flight Information cannot be reproduced in whole or in part without written permission of Dittler Bros. Inc.

Enter Departure City / Destination City in the format CITY, STate / CITY, STate TIME(s) (Times Optional) EXAMPLE: Chicago, Il / New york, Ny 0900A-0200P

****

That's simple enough. The example shows you exactly how to enter the information. Let's say you wanted a flight out of Cincinnati between 8:00 A.M. and 11:00 A.M. Here's how you would enter the information:

****

Cincinnati, Oh / Orlando, Fl   0800A–1100A <RETURN>

****

After that, you'd see:

****

DEPART	APT	ARRIVE	APT	FLIGHT	CLASS	DAYS	MEALS	PLANE	STOPS
08:00A		10:05A		ZO 0302	K	1234567		B11	0
09:29A		11:15A		DL 0505	FYBMV	1234567	S S	72S	0

****

It's typical of these scheduling services that there's a lot of information crammed into a small space. For instance, the columns of this table tell you:

- Time of departure.
- Airport of departure if there is more than one airport in the city. (By the way, you can type HELP AIRPORTS at the prompt for a listing of domestic airports and their abbreviations.)
- Time of arrival.
- Airport of arrival if there is more than one airport in the city.
- The airline and its flight number. (And for a list of the codes and the associated airlines, you can type HELP AIRLINES at the prompt.)
- The class or classes offered for the flight. (F is first-class. Y is coach. S is standard one-class service. FN is first-class night coach. QB is super coach. YN is coach-class night fare. B is an excursion option. M is supereconomy. K and V are thrift fares.
- The days of the week on which the flight operates. (Monday is 1 and Sunday is 7.)
- The meals served in flight. (B is breakfast. C is continental breakfast. L is lunch. R or S is snack. D is dinner.)
- The aircraft body type.
- The number of stops.

If no direct flights are listed for the cities you've specified, you can find the connector flights by typing CITYCON-D at the prompt. Incidentally, at this stage it's best to type the name and state of the smaller city to or from which you are traveling. Then you'll see a list of the cities which are directly connected to this city via flights. Once you make your list of direct flights, you can go back to the AIRSCHED part of the program to see the exact schedules.

For example, you could use CITYCON-D to find that there are direct flights between Huntington (WV) Tri-State Airport and Cincinnati. Then you could return to AIRSCHED-D to find the schedule of direct flights between Huntington and Cincinnati and couple that with the schedule of direct flights between Cincinnati and Orlando.

AIRSCHED-I, the international flight schedule, works basically the same way.

Using Dittler or OAG is yet another way to employ limiting "keywords" to find specific information on databases. If you think about it, it's not really different from any other keyword search. You simply are telling the database program the limits in which to work while seeking the information for you. As with any other database, the trick is to know how the program expects the information to be entered.

STAY-AT-HOME SHOPPING

Let's make this perfectly clear from the beginning: we hate to shop. Nothing's more irritating than walking two and a half miles across a shopping mall parking lot only to wander a concourse in wide-eyed amazement. We'd just as soon stay at home.

We—and you, too—now have the option of staying home and shopping at the same time with these programs.

Maybe you're a person who likes to shop. But, c'mon now. Surely there are some things you'd just as soon not spend your time shopping for. How much fun can it be comparing the prices of air conditioners at ten stores? Or how about the times you've wanted a specific book and have gone to twelve different bookstores only to find it's not available in any of them?

These are perfect opportunities for shopping via computer. To demonstrate, we'll use the mighty Comp-U-Store service.

COMP-U-STORE: AN ON-LINE SHOPPER'S DREAM

Comp-U-Store, another database offered on several services, offers electronic discount shopping for more than 50,000 brand-name products. In order to purchase them directly from Comp-U-Store, you have to be a member. The cost of membership at the time of this writing was $25 for one year and $40 for two years. That allows an unlimited number of orders during the membership period.

Wait a minute, you may be thinking. Spending hard-earned money to join a buying service without knowing whether it's a good deal? Is *that* so smart?

Well, the nice thing about Comp-U-Store is that, as a member of The Source, you can browse to your heart's content for just the cost of connect time. You can't place orders as a nonmember, but being able to window-shop gives you plenty of time to decide whether you want to join.

Comp-U-Store is located on the Shop menu of the service, which looks like this:

****

SHOPPING AT HOME <SHOP>

1 Comp-U-Store <CSTORE>
2 Musicsource Records and Tapes <MUSIC>
3 Books <BOOKS>
4 Radio Program Tapes <RADIOSOURCE>
5 Member Bulletin Boards
6 Member Publishing Catalogs <PUBLIC 1>

Enter item number, <H>elp or <Q>uit:

****

Choosing Option 1 will take you to Comp-U-Store's front door:

****

COMP-U-STORE ONLINE SHOPPING SERVICE

1 Overview/Instructions
2 Browsing for Non-Members
3 Membership Sign up
4 Shop & Order — Members Only

****

Option 1 from this menu gives you instructions about using the store; the third option provides instructions for applying for membership; Option 4 is the area you choose after you become a member and you want to place an order or bid on an item offered in an on-line auction.

Number 2 allows you to browse without signing up for membership. Here's what you'd see if you chose that option.

****

Please stand by. . .
•
•
CUCAQS V 21A(113) 21:27
•
Welcome to Comp-U-Store OnLine
•
•
Questions? Call 800–843–7777
•
All users are subject to the terms and conditions as stated below. Copyright 1982–1985 Comp-U-Card.
•
Press Return/Enter to continue or enter "SKIp" > <PRESS RETURN>

•
Type HELP for assistance at any level.
To end your visit on Comp-U-Store:
—Type EXIT to return to The Source
—Then type QUIT to return to Command Level on The Source.
Press Return/Enter to continue > <PRESS RETURN>

Main Menu

1) Help & Information
2) Member sign-up/address change
3) Comp-U-Gram
4) Scavenger Hunt
5) Comp-U-Stakes Auction
6) The Comp-U-Store
7) Best Buys (Databasement)
8) The Home Furnishing Store
9) The Comp-U-Mall
10) The Drugstore
11) THE ACTION AUCTION

****

This menu points to such special areas as Comp-U-Store's newsletter (Comp-U-Gram), the Comp-U-Stakes Auction where members can bid on special items, and the bargain Databasement where items are priced below regular discount prices. All of these sales areas can be browsed without joining Comp-U-Store. The main Comp-U-Store, where we're going, is Option 6 on this menu. Choose it and you'll see:

```
************************************************************

    **TOP**

    Enter:
    (1) Shopping & Ordering
    (2) Best Buys (Databasement)
    (3) To process a stored order
    >

************************************************************
```

Here in Comp-U-Store, the arrow (>) is your prompt.

Let's pretend we're in the market for a free-standing, 36-inch-wide electric range. We'd choose Option 1 from the menu to see:

```
************************************************************

    Enter:
    Product Type
    (LISt) for Product Listing
    >

************************************************************
```

At this point, we could simply type the word RANGE as a sort of direct command. But let's look at the list of products available by typing LIS at the > prompt. That will show you:

****

#	Categories
1	Appliances
2	Cameras & optical equip.
3	Cars, car stereo & tires
4	China, crystal and gifts
5	Silverware and cutlery
6	Luggage
7	Sports, exercise & health
8	Stereo & audio equipment
9	TV & video equipment
10	Phones & communications
11	Computers & accessories
12	Home & office furnishing
13	Tools
14	Jewelry & watches
15	Musical instruments
16	Office equip/sewing mach.

Enter A Category # >

****

If you typed 1 <RETURN> here, you'd see codes for all the appliance products Comp-U-Store stocks.

Now, these codes make Comp-U-Store seem uninviting to some people (simply because "HSAP" doesn't imply "Household Appliances" and "CKTP" generally doesn't suggest "Cooktops" to most of us. However, if you decide to become a member of this service, a handy list of the major abbreviations is among the pamphlets you receive).

Anyway, the word RANGE or RANG will work here. The program ignores the last E.

Next, the program will ask you to enter a preferred brand name. Again, by typing LIS, you can see the brands available.

Let's assume you have no preference as to brand name. You'd type NP <RETURN>. This would take you to the next prompt where you'd indicate the style you were looking for.

****

Enter:
Brand name?
(NP) No preference
(LISt) Brand listing
>NP

•

Style?
(NP) No preference
(1) Free standing
(2) With eye level oven
(3) Drop-in
>

****

We're looking for a free-standing range, so we'd type 1 <RETURN>.

****

Power source?
(NP No preference
(1) Electric
(2) Gas
>

****

Option 1 focuses on the electric ranges.

****

Width in inches?
(NP) No preference
(1) 20″ or 21″
(2) 24″
(3) 27″
(4) 30″
(5) 36″
(6) 40″
>

****

Option 5 is the 36-inch ranges. The search is narrowing. Next question:

```
***************************************************************

Cleaning?
(NP) No preference
(1) Self cleaning
(3) Continuous
(3) None
>

***************************************************************
```

No preference here. We'd type NP to see the next question:

```
***************************************************************

Microwave?
(NP) No preference
(1) Yes
(2) No
>

***************************************************************
```

Let's assume you have no preference here either.

```
***************************************************************

Ship-to State
(necessary for quote)
>

***************************************************************
```

Don't worry. You're not about to be added to an electric-range manufacturer's junk mail list. The program needs to know the state you live in so it can estimate shipping and handling charges. Just type the two-letter postal code for your state.

Next, you'll be asked to set a top limit on the amount you're willing to spend. Let's say your maximum is $1,000.

****

What's the most you will spend?
>$1000

****

That's the last prompt you'd see before the list of selected ranges is prepared. The list might appear something like this:

****

Ranges			Page 1 Of 1
#	Mfg	Model	FDC
1	MGCF	45D14	377.10
2	MGCF	45D15W	414.55
3	MGCF	45D3CLW	462.70
4	MGCF	45D3CW	446.65

Enter Selection #, or
p# for that page >

****

Sometimes the list of products is so long that there's more than one page. We see in the upper right-hand corner the total number of pages and which page you're on. If you were on page 2, typing P1 <RETURN> would take you back to the first page. (FDC means "Fully Delivered Cost," by the way.)

When you find a selection that looks interesting, just type the number and press RETURN. For example, selecting No. 1 from this list would get you the following information:

```
****************************************************************
                         Ranges
Brand:MAGIC CHEF                              Model: 45D14
List:          open
Price:         307.10        Srchg:                       8.00
With shipping and            CUBucks:                   307.10
handling:      377.10        Color:WH AV CO HW

               ------------Description------------
36" FREESTANDING MANUAL CLEANING
ELECTRIC RANGE. FEATURES SIGNAL LIGHTS,
SEAMLESS/LIFT-UP COOKTOP AND REMOVABLE
OVEN DOOR.
Want to order? (Y or N) >
****************************************************************
```

If you're just browsing without a membership, there's no way you can order, anyway, so you'd simply type N <RETURN>. That would take you back to the list of ranges where you could continue to look through the offerings.

No matter what you're shopping for, this is the type of questioning you'd go through, although the questions will be different for each item. There aren't many self-cleaning cameras available. . . .

To go back to Comp-U-Store's entry menu, you'd type EXIT at any prompt. To leave Comp-U-Store entirely, type Q at any prompt. It's as simple as that.

THE SPECIALTY SHOPS

On the same Shop menu, you'll find other stores as well. There's the professional Book Center (BOOKS) which offers electronic ordering of in-print books published or distributed in the United States.

MUSICSOURCE is a mail-order service for current albums, tapes, and cassettes. Among the features is a keyword search for records or tapes by performer, album title, or song title.

RADIOSOURCE offers cassettes of old-time radio programs which you can order on-line.

One more thing about on-line shopping. Some of your fellow Source members have goods and services for sale. You visited the POST "classified ads" section in the last chapter. In addition to that, if you type

PUBLIC 1 <RETURN> at the Command Level prompt, you'll go to the Member Publishing area. Some of those member-published features list items and services available through catalogues. Most are computer-related but there are additional catalogues on such topics as classic cars, records, etc.

A FINAL WORD

When you're dealing with on-line merchants, you most likely are not dealing with The Source directly. That means that most of them want payment via check or credit card to their own companies. In other words, the cost of the item purchased usually can't be added to your Source bill. This is generally explained in detail in the order area for the shopping service.

Think of The Source as the mall that rents the space to these merchants, not as the merchant itself. But since The Source is renting the store building, The Source would like to know if you have any difficulties with any of the merchants. Call The Source customer-service number if you do.

LETTING THE SYSTEM HELP YOU

Finally, don't be afraid to continue experimenting in these new areas with the system tools you've been learning. For instance, just because you learned about COMO in connection with POST doesn't mean you can't use it here.

Suppose, for example, you and some friends were communicating through SourceMail about meeting in a city somewhere in between your present locations. You might decide to each share part of the planning, such as the searching of the flight schedules, hotel and restaurant guides, etc.

You now have enough background on the system to be able to share this information in rather sophisticated style. You could:

1. Open a COMO file in order to put all this research in your User File Directory.
2. Visit the hotel guide and find the accommodations that best meet your needs.
3. Exit the hotel guide and turn off the COMO (entering COMO –E at the –> prompt, remember).
4. Edit the captured file to remove all but the last part—the summaries of the hotels themselves—and then save the edited version.

5. Send the findings to your friends via SourceMail, using the dot-LOAD command to pick up a copy of the file from your UFD. Then delete the file after it's been mailed.

Ah, it's just one more example of making it easier on yourself—by making The Source work harder.

CHAPTER 9

The Games Source Users Play

The Source can play for you as hard as it can work for you. It has some of the classic computer games on-line. But before we look at them, let's clear the air about something.

The personal computer community has become polarized between the Plays and Play-nots. There are some who think that games are a noble and legitimate exercise for their machines. Many think the games are even better if they're on-line, because players can compete against a giant computer, one that's doing a hundred other things at the same time it's challenging you.

The other camp thinks of game playing as poor use of a good microprocessor.

Well, we choose forthrightly to skirt the issue. As a rule, we don't discuss politics, religion, or recreational computing.

Of course, had games not been so important in our culture, our language would have been the poorer for it. Government would then, of course, come to a standstill—parliamentary arguments stymied, politicians stalemated without a single verbal trump card with which to turn the trick and . . .

Ahem. Excuse that. We got a little carried away there.

To avoid taking sides in this debate, we're suggesting that the following is an optional on-line tour. If you're curious about the system's game collection, come along and we'll look at it together. If you're not, just peruse the next few pages from the sidelines. No one's suggesting you'll be left behind the eight ball if you don't touch all the bases—stop that!

This chapter is divided into three parts:

First, we'll take a very quick on-line tour to see a few of the games firsthand.

After we log off, we'll catalogue some of The Source's major games in case you want to explore them further on your own. Incidentally, some of these have been real mainstays of microcomputing history. If you got your first computer five or more years ago, you might have had some of them—Star Trek, Adventure—in the original software library with your Apple II, your TRS Model I, your PET. However, if you're a more recent convert, you might want to visit some of these quaint treasures in the interest of studying your computer heritage.

Finally, we'll be back with tips on how to make your own fun on-line. Using some of the Communication Services we've already seen, you can find other chess players for an impromptu tournament or other dial-up friends who just like to talk about contemporary computer gaming.

OUR FIRST MOVE

Log on for our quick spin through the game arcade. Turn CHAT -OFF as usual.

There's no need for us to go through the list of games on the menu. We did that in Chapter 3, if you'll recall, when we went all the way through the Games menu to the Pits which, appropriately, is at the end of the longest menu on The Source.

Once you're on-line, go to Command Level.

We're taking a little different approach in this chapter. Obviously, there's no way we can show you how most of these games are going to go for you. All of them are different. But we're going to guide you in some general instructions and let you take it from there. You'll be on your own, but we'll be nearby to call out instructions.

Now, at the command prompt, type:

****

-> PLAY SUPERBJ <RETURN>

****

This will take you to the Super Blackjack game, a computerized version of an old favorite.

If you're not familiar with the game of blackjack, here are the rules:

the dealer (in this case, the computer) gives you cards one at a time. Numbered cards count the number which is on them. Face cards count 10. The ace counts 1 or 11. Your job is to keep accepting cards to get as close to 21 points without going over. Meanwhile, if the dealer gets closer to 21 than you without going over, the dealer wins. You'll be able to bet any even amount between 2 and 500 dollars. The good news is that if you lose, the amount isn't tacked on to your Source bill. The bad news is if you win, your winnings won't be paid to you either. This game's just for fun.

When the system asks how many are playing, type 1 <RETURN> and type in your first name at the next prompt while the dealer shuffles the cards.

Right now, all you need to know is that by typing HIT <RETURN>, you're telling the computer dealer you want another card. Typing PASS <RETURN> means you don't want another card.

There are many more commands—even some that let you cheat and take a chance on being caught. You can check them out by typing HELP COMMANDS <RETURN> at the prompt.

Go ahead and play a couple of hands to get a feel of it. When you're finished, Q <RETURN> will end the session. Because you're pushing away from the table so soon, the computer may send you a little nasty message. Like so many dealers, he's a little touchy.

ROUND TWO

Back at the -> prompt.

As you'd expect, Command Level is the door that leads to all the games, as long as you know their direct addresses. If you should want to see the Games menu itself, you can type GAMES <RETURN> at the prompt.

Now, let's try a few moves in the on-line checkers game. Type:

****

-> PLAY CHECKERS <RETURN>

****

You'll see a short introduction, then you'll be asked the IQ level at which you want to play. Type 1 <RETURN>, since this is the level at which the computer plays the fastest and simplest game. If you selected level 5, the computer might take up to two minutes to make a move.

The system will draw the board, like this:

```
*****************************************************************

   FORCED JUMP MODE:     IQ = 1

    1    2    3    4    5    6    7    8
   ###   ×   ###   ×   ###   ×   ###   ×   1
    ×   ###   ×   ###   ×   ###   ×   ### 2
   ###   ×   ###   ×   ###   ×   ###   ×   3
        ###       ###       ###       ### 4
   ###       ###       ###       ###       5
    0   ###   0   ###   0   ###   0   ### 6
   ###   0   ###   0   ###   0   ###   0   7
    0   ###   0   ###   0   ###   0   ### 8

*****************************************************************
```

The computer always plays the X pieces; you always play the O pieces. The blanks are the spaces where the pieces move. The ### characters show where the checkers can't move.

You make a move by typing four numbers. The numbers across the top are columns and the numbers down the right side of the display are the rows. You type the column-row of the piece you are going to move and then the column-row of the space where you want to move it. So if you wanted to move the O piece in column 7, row 6 to the space that is in column 6, row 5, you'd type:

```
*****************************************************************

   7665 <RETURN>

*****************************************************************
```

After you move, the board is redrawn by the system.

```
*****************************************************************

   FORCED JUMP MODE:      IQ = 1

    1    2    3    4    5    6    7    8
   ###  X   ###  X   ###  X   ###  X    1
    X   ###  X   ###  X   ###  X   ### 2
   ###  X   ###  X   ###  X   ###  X    3
        ###       ###       ###       ### 4
   ###       ###       ###  0   ###       5
    0   ###  0   ###  0   ###       ### 6
   ###  0   ###  0   ###  0   ###  0    7
    0   ###  0   ###  0   ###  0   ### 8

*****************************************************************
```

If the computer moves the same way it did when we tried that move, it will display this:

```
*****************************************************************

   Hmmm.
   Considering a move.
   Thinking it over.

   The SOURCE moves from: 6 3
   to position: 5 4

   FORCED JUMP MODE:      IQ = 1

    1    2    3    4    5    6    7    8
   ###  X   ###  X   ###  X   ###  X    1
    X   ###  X   ###  X   ###  X   ### 2
   ###  X   ###  X   ###       ###  X    3
        ###       ###  X   ###       ### 4
   ###       ###       ###  0   ###       5
    0   ###  0   ###  0   ###       ### 6
   ###  0   ###  0   ###  0   ###  0    7
    0   ###  0   ###  0   ###  0   ### 8

   Enter your move (4 NUMBERS) ->

*****************************************************************
```

And so it goes. You move, then the computer moves, till the end of the game. Unlike the Blackjack dealer, the checker player won't tolerate cheating—the computer won't let you make an illegal move. Notice this is a game in which you have to "take your jumps," as Granddad used to say.

When you're ready to leave, simply type as your next move

****

9999 <RETURN>

****

and you'll be whisked back to the –> prompt.

GOING FOR THE GOLD

Text adventure games are as much a part of microcomputer history as the machines themselves, and The Source has several stashed away. Our last stop on the tour is a good example. At the prompt, type:

****

–> PLAY DUNGEON <RETURN>

****

The game begins immediately by describing a white house with a mailbox in front. For those not familiar with adventure games, this is the perfect invitation to open the mailbox. So, type OPEN MAILBOX <RETURN> at the question mark prompt.

A leaflet inside. Try READ LEAFLET <RETURN>.

Hmmm. Maybe not much help in solving the puzzle, but at least it gives you a historical footnote on the game which you may find familiar.

At the next prompt, type NORTH (or N) <RETURN>. (When typing directional commands, you don't need to type GO NORTH. The direction or N, S, E, W is adequate.)

Well, there seems to be nothing here. Type EAST (or E) <RETURN>.

A window, slightly ajar. Here's an opportunity. Type OPEN WINDOW <RETURN>.

The window is open. How about—GO IN <RETURN>? This brings you into the kitchen with lots of opportunities. A passage leads west. A staircase leads upward. A small window is open to the east. There's a sack that smells of hot peppers and over here's a glass bottle containing water.

This is where play becomes interesting. Which way to go? Do you drink the water? Do you, perhaps, carry the water with you on your journey? Do you grab the sack and open it?

Play a little longer if you wish, but remember that adventure games like this are habit-forming. When you're ready to stop, type QUIT <RETURN> at the prompt. (You'll be told how many points you have scored out of a possible 585. The program will ask if you really want to leave. Type YES <RETURN> to get back to the -> prompt.)

We'll leave you in the kitchen. When you've wandered the twisty passages to your heart's content, meet us off-line and we'll summarize The Source's game offerings.

THE GAMES COMPENDIUM

As promised, let's take a closer look at the major games available on The Source.

Now, frankly, none of the games you'll see here can match the top-of-the-line entertainment software being sold these days. That's because graphics and sound and speed have become so much a part of contemporary game crafting that all of the dial-up services are hard-pressed to keep up, considering the nature of computer communications itself. Also, The Source has unabashedly put its development efforts into other areas—like the business and financial features we'll be seeing later, the searching utilities we've been using all along, and the communication services that we'll continue exploring in the next chapter.

That doesn't mean The Source game arcade is without value. Far from it.

As we've said, many of the games are truly historic in this field. Also, many have a special interest. For instance, some people just can't pass by a puzzle without trying to solve it. In computer gaming, text adventures still are at the top of the heap and The Source has some fine examples of the genre. Finally, on-line games sometimes are a wonderful way to introduce children (and technophobes) to computers and on-line services. They're often put more at ease about computers when they see that the system knows how to play as well as work.

Most of the on-line games have help files—a brief article that is an introduction to the game—available from the Command Level prompt.

For instance, to get to the instructions for Super Blackjack from the -> prompt, you can type HELP SUPERBJ <RETURN>.

Most also have instructions within the game program itself. So, if you get involved in a game, you generally can summon help without stopping the play.

Some games allow you to save your place midway through the play and return later. When you do that, you are creating a file in your User File Directory. Remember, you'll be charged an additional storage fee while that file is there. That's why it's a good idea to periodically check the files in your UFD by typing LD <RETURN> at the -> prompt. If you change your mind and decide *not* to resume the game, you then can remove the file with the DEL command we've shown you.

Here are some of the games available on The Source, including a description of each and its direct address. As always, be prepared for changes—additions and deletions—as The Source continues to evolve.

Blackdragon <PLAY BLACKDRAGON>

Blackdragon is a fantasy and role-playing game in which you accumulate treasure, convert gold into experience points, and become strong enough to conquer the Arch Demon on the tenth level of the maze. You can find useful objects and magic functions in the labyrinth or you can purchase these things from vendors if you have enough gold.

It's a good idea to map the maze to help you through each level should you meet misfortune and have to try again.

It generally will require several sessions to conquer all the levels. If you quit or are killed while in the maze, your character is lost forever. If, however, you safely exit a level of the maze, your character is saved in a file called ".MAZE" in your file area. Once in the maze, you move N, S, E, or W for the points on the compass. Or you can teleport directly through the maze—but you've got to learn how from experience.

Among the "monsters" awaiting you is one called "SOURCE programmer." Veterans say he's a real pussy cat, though, and you know veterans don't lie.

Super Blackjack <PLAY SUPERBJ>

This game of blackjack allows you several options. You can have one to five players per game. You can play Las Vegas, Atlantic City, or "friendly" game rules. You can practice card-counting techniques. You can play with from one to ten decks of cards. You can play with a "fixed" deck. You can mark the cards. You can peek at the next card or all the cards. In addition, Super Blackjack maintains a list of the top ten winners in the game from around the system and you can try to make the honor roll.

Once in the game, you can type HELP RULES or HELP COMMANDS at the first prompt to learn all the rules of this game.

Castlequest <PLAY CASTLEQUEST>

This is an interactive adventure game. The object is to find the master of the castle and kill him while accumulating as many treasures as possible. You direct the game with one- or two-word commands—LOOK, TAKE, DROP, UP, DOWN. If you are in the middle of a game and want to leave, you can type SAVE <RETURN>. When you return, type RESTORE <RETURN> at the first prompt to continue where you left off.

To help you along, you can ask for tips by typing HINT <RETURN> and the name of the object you are asking about. You can get help about a room if you type HELP ROOM when you enter. Remember that some objects that appear to be worthless may be valuable later in the game, so don't pass something up just because it doesn't appear to have value.

We could find no on-line help for Castlequest available at the command prompt. However, you can get complete instructions by answering Y to the question "Do you want instructions?" once you are in the Castlequest game area.

Adventure <PLAY ADVENTURE>

Once you enter this game, you explore Colossal Cave to find its hidden treasures. The object is to collect as many treasures as you can without getting killed by the dangers that lurk in the cave. Each treasure has a point value. The maximum is 350 points. As with many adventure games, most commands here are one or two words. To get out of the game, type QUIT or tap the BREAK or CONTROL P keys. If you want to stop the game and continue it at a later time, type SAVE at a prompt. When you come back, type RESTORE at the first prompt.

Adventure 550 <PLAY ADV550>

This is a more complicated version of the previous adventure game. Everything that's in the 350-point version is in this one and more. The maximum number of points to score is 550 points. There's no help for this game by typing HELP ADV550 at the prompt. However, there's help once you're in the game program.

Hangman <PLAY HANGMAN>

This is the computer version of an old game. You try to find the word a computer selects by guessing the letters in the word before your effigy is hanged. If you make eight wrong guesses, the noose tightens. The computer begins by printing as many dashes as there are letters in the secret word. You select the letters you think are in the word, one at a time, by typing the letter (with or without a carriage return). If you select a letter in the word, the dash where it's supposed to be will be replaced by the letter. You can exit the program and get back to the Command Level by typing BREAK or CONTROL P.

Checkers <PLAY CHECKERS>

You saw the checkerboard of this game during the tour. The X pieces belong to the computer; the O pieces are yours. The board is divided into an x-axis (left to right) and a y-axis (top to bottom). Except for double jumps, you move by typing four numbers without spaces between them. The first two numbers are the x-y coordinates of the piece you want to move and the last two numbers are the x-y coordinates of the space where you want to move. You can't move into any of the spaces represented by the three number signs.

An example of a good opening move might be 7685. If you made that move, the board would be redrawn with the move indicated and the computer would take its move. If your move is illegal, the word **ERROR** will appear. Check to see what is illegal about the move you made.

In double jumps, enter the four numbers for the coordinates of the first jump, then the game board will be redrawn, then you need only enter the coordinates of where the checker will land after the second jump is made.

You can ask for an easy game by typing 1 when it asks for the IQ level at which you want to play. For harder games, type 2, 3, 4, or 5. At an IQ level of 5, the computer may take one or two minutes to think.

To end the game, type 9999 when it's your move.

Dungeon <PLAY DUNGEON>

Dungeon is another adventure game, one we met on the tour. You seek treasures while trying to avoid the seemingly endless dangers in the dungeon maze.

Once in the game, type HELP and INFO for instructions or you can type HELP DUNGEON at the command prompt.

Printwiz <PLAY PRINTWIZ>

Printwiz isn't a game. It's a poster-printing program and if you have a printer, it'll work for you. You enter the message you want printed, and the message is printed out sideways on your printer so that the end result will look like a banner which is suitable for hanging.

It's always best to turn CHAT off before doing a poster with Printwiz. As far as making a poster, it's quite easy, although there are several commands for tailoring your printout.

PRINT is the command to send to an 80-column printer. PRINT LARGE is the command for a printout on a 132-column printer and PRINT SMALL is the command for a 40-column printer. The CHARACTER command can be used to print "negative" characters or a character outline. HELP TEACH at the Printwiz program main prompt will give you a complete rundown on the many options available.

Terminal Trivia <PLAY TRIVIA>

This is the on-line version of the game that took America by storm a few years ago. This one has a few extra options, however. You can place bets in this game just to make it interesting.

Five persons or five teams at a time can play. The winner will be the person or team with the most points and right answers.

New topics are occasionally added. Players with the highest recent scores often are listed in the Trivia Game area to give you something to shoot for. Commands include the following:

(C)AT	—selects a question category
(H)ELP	—describes terminal trivia
(T)YPE	—chooses type of play
(S)TATUS	—shows game status
(W)IN	—displays winner's list
(P)LAY	—starts playing TRIVIA
(R)ESTART	—restarts game from scratch
(Q)UIT	—exits TERMINAL TRIVIA

Explore <PLAY EXPLORE>

This adventure game allows you to explore Mystic Cave for treasures while attempting to avoid all the dangers inside. There is no on-line help for this game from the Command Level prompt. However, if you type Y for Yes when asked if you want instructions once you are in the program, you'll get a complete listing of the game's scenario.

I Ching <PLAY ICHING>

This is an interactive computerized version of I Ching (pronounced Yee Jing), an ancient Chinese method for making decisions, determining the most probable course of future events, and choosing the best course of action.

Using I Ching involves a question-and-answer procedure. This version of I Ching is useful for neophytes, since it contains complete information about the theory of I Ching in several on-line text files in the game program itself. There are tutorials in such things as an overview of the I Ching method, how to compose your question, and how to interpret a reading. These are all included in the TEACH section of the program. Type Q from the Command Level to leave the game.

IQ Test <PLAY IQTEST>

The game board in this program looks like this:

```
*************************************************************

              1     2     3     4     5
                 6     7     8     9
                   10    11    12
                      13    14
                         15

*************************************************************
```

You may remember this one from your youth. Picture this triangle to be a board of pegs. You can start by removing one of the pegs anywhere on the board. You jump the pegs one at a time into an open hole (which is shown by a period), removing the peg you jump.

The object is to remove as many pegs in this manner as you can. Sound simple? It isn't. After each jump, the computer redisplays the board. Use BREAK or CONTROL P to leave the game.

Backgammon <PLAY BACKGAMMON>

The backgammon board you see on The Source may confuse you at first. Here it is:

```
*****************************************************************

              R             W   R             W
              R             W   R             W
              R     R       W   R       W     W
    W         R     R       W   R       W     W         R
    W         R     R       W   R       W     W         R
+   + + + + + +   + + + + + +   + + + + + +   + + + + + +   +
1   2 3 4 5 6 7   8 9 0 1 2 3   4 5 6 7 8 9   0 1 2 3 4 5   6

*****************************************************************
```

It looks different because your opponent's side has been rotated and placed to the left of the screen. A plus sign represents a point on the board. There are two points by themselves on the left and right. They are occupied by men after they have been hit. The computer plays red (R) and moves from right to left. You use the white men (W) and move from left to right.

The numbers below the points are used when moving your men. The points are numbered from 1 to 26. You must mentally add the tens digit when appropriate. Moves are made by specifying the "From" point and the roll of the die.

There's no file called HELP BACKGAMMON available from the -> prompt. Instructions are available once you're in the game program, however. When asked if you want instructions, type Y for Yes.

Biorhythms <PLAY BIORHYTHMS>

This is another nongame in the game area. You provide your birthdate and the number of days you want charted, and the program gives you your biorhythm chart for that period. If you're going to chart more than a few days, it's a good idea to get a printout. The symbols on the graph are:

:	Median Line	(0-Line)
P	Physical Cycle	(23-day)
S	Sensitivity Cycle	(28-day)
I	Intellectual Cycle	(33-day)

Not only does this program provide a chart, but if you want a brief history of the biorhythm theory, answer N for No when the program asks if you know what a biorhythm is.

Pits <PLAY PITS>

Yet another interactive adventure game. Once more you descend into the Great Pit to retrieve treasure and face danger. One- and two-word commands are the order of the day. The first time you enter, type YES at the first prompt to get information about the game and type INFO at any prompt to get information on how you can customize the game. CONTROL P or BREAK will get you out of the game.

*Trek <PLAY *TREK>

This is an adaptation of a Star Trek game. Part of the game is to learn what to do and how to do it. The scenario is this:

The USS *Enterprise* is on patrol in the Arcturus Sector of the galaxy near Gamma Triskellion-6. Your mission is to destroy Klingons. You must destroy ten Klingon battle cruisers in twenty-six stardates. There are five starbases in the galaxy. Type H <RETURN> for a list of commands.

Othello <PLAY OTHELLO>

This computerized version of the popular board game is a test of skill and strategy in which you attempt to capture your opponent's pieces by "outflanking" them while your opponent tries to do the same to your pieces.

Othello is played on an 8 $\times$ 8 checkerboard. The rows are numbered 1 to 8 and the columns A to H. This computerized game uses letters X and O for pieces. The initial board looks like this:

```
**************************************************************

     A B C D E F G H
  1  — — — — — — — —
  2  — — — — — — — —
  3  — — — — — — — —
  4  — — — O X — — —
  5  — — — X O — — —
  6  — — — — — — — —
  7  — — — — — — — —
  8  — — — — — — — —

**************************************************************
```

When you start playing Othello, a file called OTHELLO.SAVED is created in your file area and will remain there until the game ends normally or you delete it yourself. A game ended by BREAK or CONTROL P will not automatically delete the file. A game ended with a Q <RETURN> will automatically delete the file.

WRAPPING UP

First, a word about the cost of on-line games. Maybe you're asking yourself why you would want to play on The Source. Isn't that rather expensive?

Maybe, and maybe not. For example, once you solve an off-line adventure game that you purchased for $40, how many times do you play it again? In fact, after purchasing most computer games and playing them a few times, they tend to gather dust. You might think of playing the games on-line as "leasing" them for a time until the interest wears off. Depending on how you play, you might find you save money in the long run.

Some thoughts on making your own fun:

Even if none of these games quite ring the bell, a sporting person still can find a game or two on-line by using some imagination and the electronic resources we've learned.

For instance, Charlie has played a number of on-line chess games, sending and receiving notation of the moves via electronic mail. It's just like traditional postal chess, only faster. And you could use the CHAT feature for a game or two, particularly for a "rapid transit" game in which less time is spent between moves.

Most other strategy games can be recorded, and anything that can be written down can, of course, be delivered through SourceMail. Regular gamesters on the system are always coming up with clever schemes for annotating various games.

The point is, if you're interested in games, you should get to know the others on-line with similar interests. We'd suggest you start with the Member Directory and search the <IN>terest field for, say, CHESS, or CHECKERS, or GAMES.

After you've found some likely candidates, drop them a line through SMail, maybe a bulk letter to all of them to see if they'd like to form a circle of correspondence. Maybe they already have and you'll be invited to join, too.

Remember also that there's a category on the POST bulletin board devoted to recreation. Try POST S GAMES at the –> prompt from time to time and see what you come up with. If you find some fellow gamesters, perhaps you can move over to SourceMail or CHAT for some play.

Finally, there's one other area of recreational computing we haven't addressed yet. Many of the public-domain programs available around the system are games. Many hobbyist programmers write games to share with others on-line.

However, this opens up a new topic: to get at them you have to know some things about transmitting and receiving files to and from the system. Well, it's a little late in the evening to be broaching such a broad new subject. So, hold that thought—we have a whole chapter coming up on that subject.

CHAPTER 10

The PARTI Experience

If there's one thing that sets The Source apart from other information services, it's PARTICIPATE (nicknamed PARTI), an impressive computer conferencing program developed by Participation Systems, Inc., of Winchester, Mass. With it, you can read messages and post your own in scores of different ongoing discussions about topics ranging from news and gossip to ideas from the business world to tips on solving tough computer problems or losing weight.

Perhaps, if you have been a Source user for some time and traveled The Source a little on your own, you've already logged into the PARTICIPATE area and, knowing little about it, immediately got lost. Some decide to return a few weeks later to find everything has changed, or at least it seems to have changed. That multiplies the confusion geometrically.

You are not alone. Some people say they learn the PARTI system quickly and easily; it makes all the sense in the world to them. Others, maybe the majority, say they have trouble with the concept.

Well, help has arrived. In this chapter and the next one we show you how to become familiar with the basics of PARTI in a short time. You'll learn:

• Where to find important, helpful and entertaining discussions (or "topics"), how to read the interlocking messages in these topics and how to add your two cents' worth.

• How to save money in PARTI by reading, scanning and writing notes quickly.

• How to use PARTI's own electronic message system, a way to send private notes to others who use PARTI on The Source.

• How to search messages for subjects that interest you.

• Where to find help—lots and lots of help, ranging from an index of ongoing conferences to personal attention from people who just love to answer your questions.

But what we hope you learn above all else is that the structure of PARTI doesn't change. The framework for the ongoing discussion topics doesn't change. But the topics change daily—even from minute to minute at times.

Why do we want you to learn that? Because, we believe, that's where a lot of the confusion about PARTI is generated. Once you become familiar with that important PARTI framework, the rest will come easily.

AN OVERVIEW OF THE ELEPHANT

Remember the story of the blind men and the elephant? Well, this old tale applies to PARTI on The Source. If you'll recall, the story involves a group of blind men who each had hold of a different part of an elephant. The one who grabbed the trunk was sure he was touching a snake, for example. None of them guessed they had hold of an elephant because none of them got an overview.

Well, in talking to people who have gone into PARTI, we've decided that a lot of them fail to get an overview of this electronic elephant which has so many current and potential functions. In their brief tour of PARTI, all they see, perhaps, is one or two discussion topics and perhaps those topics are instructive rather than informative or entertaining. Perhaps they get the impression—as we did when we were new to PARTI—that the electronic conferences which they find are actually the structure of the PARTI system. They aren't. They are the words—the data, if you wish—built around the conferencing structure that PARTI has built.

Here's an example. Consider how a computer program uses data in a program such as a spreadsheet or a database. The program itself provides the structure where the numbers are entered. Take away the numbers and the structure is still there although it is more or less invisible. Replace the numbers with new numbers and the structure remains the same. The framework is what holds it all together, no matter what numbers you put in it.

So it is with PARTI. PARTI provides the *structure* for some very sophisticated electronic conferencing. The users provide the *data*. They fill in the framework to suit their own needs.

This first chapter about PARTI will be a close inspection and overview of the framework. We're not even going to think yet about how to get to PARTI on The Source. Instead, we'll concentrate first on the PARTI concept and go into considerable detail about the commands you'll use to navigate that framework. Trust us, having the overview firmly in mind will make it easier in the next chapter, when we look at how to reach the PARTI system and talk about some of the ways PARTI users have decorated that framework to suit their own needs. And along the way this long chapter should whet your appetite.

ONE BIG CONFERENCE HALL

When trying to get the concept of the PARTI system in mind, we found that it was a lot easier if we thought of PARTI as a big conference hall where there's lots of activity going on at the same time.

There's a reception area that you see upon entering and where you get a rundown on the topics being covered in the conferences that interest you. That reception area also contains an internal post office where other people in the hall can leave you private mail and where you can answer their private notes on the spot.

And, finally, there's a big bulletin board where announcements of new conferences are posted and invitations are offered to join in discussions of these new topics.

These "rooms" in the conference hall never change. Conferences are always beginning and ending, the number of letters in your mailbox and the messages those letters contain change constantly. The conference topics change all the time, but the rooms—the structure itself—don't change.

Once in the PARTI conference hall, you can move freely among the conferences you are invited to attend. You can go into any of these conferences and find out what's going on by reading previous "topics" or messages left by others as part of the conference. You can add your two cents in response to any of these messages and it will become part of the conference record. If you like, you can put your name on the membership roster of the particular conference and you'll be told when new messages are added to the topic being discussed.

Perhaps, while attending this conference, you will see the need to form a kind of subcommittee to discuss a related issue. In that case, you can alert those in the room to the fact that you're going off into an adjacent area and starting a subconference. You can invite as many of the original conferees as you like.

Maybe, after wandering in and out of conferences till you've seen most of them, you will decide to start a brand new conference on a topic that interests you. You have the chance to announce your intentions on the "big board" in front of the hall inviting a few of your friends or everyone in the hall to join. Once you originate such a conference, you have considerable control over it because you originated it.

Your conference can go on and on until you've discussed the topic as much as you, or anyone, wants to thrash it out. And when it ends, you see the real beauty of this form of conferencing. Instead of leaving the hall full of scrap paper and empty coffee cups and with an empty feeling that what went on there has been lost, you have a complete written record of what transpired in the form of all those notes that were posted.

NO LIGHTS, NO CAMERA, JUST ACTION

The heart of the PARTI system is the ACTION prompt, just as the heart of The Source is the command prompt, which we've seen throughout our on-line journeys together.

In its simplest form, the ACTION prompt looks like this:

```
************************************************************

ACTION->

************************************************************
```

and we'll be showing you how to find it.

Unlike the command prompt on The Source, however, the ACTION prompt changes according to where you are in the PARTI conference hall and what you're doing at the time. Think of the ACTION prompt as more than just a place where you enter a command. Depending on what you're doing at the time, words follow the ACTION prompt. Sometimes those extra words are designed to remind you of what topic you're looking at and where you are in a string of related messages. At other times, words appear in parentheses to give you some idea of what you might want to do next. We'll see examples of both as we go along.

No matter what words follow the ACTION prompt, you can use almost any of the navigating commands in the system from any ACTION prompt, no matter what the prompt looks like.

THE BIG READ

Now for some really good news: You could spend all your time on PARTI using a single command. It's called READ. You use it to . . . well, read—that is, to call up notes left in a topical discussion and READ them or to call up private messages left for you by other PARTI users and READ them.

Let's say you've heard about a public conference called "AMERICAN HOT TUBS" in which participants discuss the various kinds of hot tubs available and the benefits and drawbacks to owning them.

The exact name of the public conference, "AMERICAN HOT TUBS," is all you need to know to READ it.

At any of the ACTION prompts, you simply type:

****

ACTION–>READ "AMERICAN HOT TUBS"<RETURN>

****

All you have to remember is to put the name of the conference in quotation marks. All conferences you attend, whether they are public or private, have short names and when calling for any sort of action on these conferences, the name of the conference must be put in quotation marks.

What you'll see after requesting a READ of "AMERICAN HOT TUBS" is not the entire conference. What you'll see is some important information about the conference. It might look something like this:

****

"AMERICAN HOT TUBS" by DAVE, March 1, 1985 at 13:04 Eastern about A GUIDE TO WATERY PLEASURE (250 characters and 8 notes)

Welcome, friend to this conference about a fast-growing form of home entertainment and relaxation: the hot tub. With so many brands to choose from and so many varieties of tubs offered, this topic has been set up to share consumer experiences. So join the fun! The water's fine!

ACTION on "AMERICAN HOT TUBS" 0 (of 8)–>(Join)

****

Notice that the ACTION prompt has changed. Before it said simply ACTION–>. Now is says ACTION on "AMERICAN HOT TUBS" 0 (of 8)–>(Join). This is what we meant when we said that the ACTION prompt is a chameleon of sorts. It changes to fit the situation and it's for your convenience.

Let's take it from the top and inspect this note which is called a topic "opener." Remember, we said that the structure of PARTI stays the same, but the words change. So it is with "openers." Their structure is always the same, although the words change.

First comes the name of the conference, in quotation marks, just as you need to remember it. Next comes the "PARTI name" of the person who organized it. (More about PARTI names later.) Then comes a short description of what the conference is about. It's always preceded by the word "about." Then there's the date the opener was written on the system. Next comes an advisory of the number of characters the opening note contains and a tally of how many notes are connected to the discussion.

Finally, there's the text of the opener itself—a more detailed, but usually still rather brief, discussion of what the organizer hopes to get out of this discussion.

At the end you see the new form of the ACTION prompt. It tells you where you are, then it makes a suggestion as to what you might want to do next.

The information contained in the ACTION prompt is important. It tells you that you are temporarily connected to the conference called "AMERICAN HOT TUBS," and that you are at the end of note No. 0 in this discussion which contains eight more notes—No. 1 through No. 8.

How can there be a note numbered zero? Hmmm, we won't discuss the philosophy here; you can study nihilism on your own time. Just be satisfied with the knowledge that the "opener" of every discussion started on PARTI is No. 0 and it's just as much of a number as any other.

The word in parentheses, in this case "(Join)," changes according to what you're doing at the moment. It's merely a suggestion. The system is trying to second guess what you want to do at this point. When you see that parenthetical command, all you have to do to execute it is press <RETURN>. On the other hand, if you have some other command in mind, you simply type it in. In other words, parenthetical commands, like "(Join)," are just timesavers.

By this time, you may have decided that you'd like to automatically receive all the future notes that might be added to this "AMERICAN HOT TUBS" topic. The only way to do that is to JOIN the topic. To join, simply press <RETURN> to execute the "(Join)" that you see at the ACTION prompt. Or, type JOIN. Since, by reading the opener, you are "connected" to the "AMERICAN HOT TUBS" conference, the system "assumes" it's the hot tubs conference you're indicating with the JOIN command.

KNOWING YOUR "NEXTS"

After joining, the system will send you confirmation of your status, like this:

****

You are now a member of "AMERICAN HOT TUBS".
Type READ followed by appropriate note numbers to read existing notes.

ACTION on "AMERICAN HOT TUBS" 0 (of 8)-> (Next)

****

That looks straightforward enough. The prompt says you are still at the bottom of note No. 0 and there are still eight more to read. But what you see next—the word "next"—is a little tricky.

You might assume that in this case it means "Read the next note in this topic." It doesn't. You will see the suggestion "next" come up several places in the PARTI structure. And each time it can mean something different.

Generally speaking, it means to read the next note in the sequence of notes you have requested. But when you type READ "AMERICAN HOT TUBS" from the ACTION prompt, the system assumes that all you want to see is the conference's opener, note No. 0 of the series. As a result, entering the next command at this point would take you back to the preceding ACTION prompt. In other words, you would be "disconnected" from "AMERICAN HOT TUBS" and taken back to a more general ACTION prompt where you aren't connected to the conference any more.

But what if you want to catch up on what has happened in the discussion so far? The prompt is telling you that you are still connected to the "AMERICAN HOT TUBS" conference, you have just read the opener and there are eight more messages to read.

At this point, you have several options and all of them will be covered later. Right now, you probably would use the READ command followed by the range of numbers you want to read, separated by a hyphen. If you want to read them all, type READ 1-8, like this:

****

ACTION on "AMERICAN HOT TUBS" 0 (of 8)->(next) READ 1-8 <RETURN>

****

You're telling PARTI you want to read notes 1 through 8 sequentially. When you press <RETURN>, you'll start seeing the first in your requested sequence, note No. 1, which might look something like this:

****

1 (of 8) CHARLIE March 1, 1985 at 14:03 Eastern (450 characters)

I'm glad someone started this discussion. My wife and I have spent the past five weeks listening to hot tub salesmen tell us why their units are the best and we're getting pretty upset. Besides, we've tested out so many hot tubs our skin is permanently wrinkled. Anyone out there who can tell me how to get these wrinkles out of my skin?. . . .

****

If the note is a long one, it will stop after about 24 lines and a "-More-" prompt will appear. As with other areas of The Source, this allows you to stop and read a screen before you go on. To continue, simply press <RETURN>. When the note ends, you should see something like this:

****

ACTION on "AMERICAN HOT TUBS" 1 (of 8)-> (Next)

****

Ahh! The ACTION prompt has changed again to reflect the progress you've made. You're still connected to "AMERICAN HOT TUBS" but you're at the bottom of note No. 1 now. And now the "Next" prompt has a different meaning.

It does?

Yes. Remember you gave the command to read notes 1-8? Well, the system remembers that. So "Next" in this case now means "Do the next part of the command I gave you earlier." Thus, by pressing <RETURN> you'll execute the parenthetical command ("Next") and see note No. 2. At the bottom of that note, you'll see:

****

ACTION on "AMERICAN HOT TUBS" 2 (of 8)-> (Next)

****

By continuing to press the <RETURN> key, the system will let the "AMERICAN HOT TUBS" saga unfold, one note at a time, all the way to the end. When it executes the entire READ 1-8 command you gave it, you'll see:

****

ACTION on "AMERICAN HOT TUBS" 8 (of 8)–> (Next)

****

And note that "Next" has changed meanings again. Since the READ 1-8 command has been exhausted, pressing <RETURN> now will take you back to a more general ACTION prompt—usually the one you saw before you started reading the current conference.

MORE ON READING NOTES

Reading notes is what you'll spend most of your time doing on PARTI, no doubt, so PARTI has made it as easy as possible for you to do it. You can read notes in several ways and from any of the ACTION prompts you might see. For example, let's say you wanted to re-read the notes in the "AMERICAN HOT TUBS" conference, just as you were ready to leave PARTI. The ACTION prompt might display ACTION–>(Quit) which means that if you hit the <RETURN> key you would be taken out of PARTI.

From this ACTION prompt, you could type the following to get back to the "AMERICAN HOT TUBS" topic:

****

ACTION–>(Quit) READ "AMERICAN HOT TUBS" 0-8 <RETURN>

****

In other words, "Hey, PARTI, I'm not ready to Quit. Instead, find this topic and start displaying the notes, from the opener to the last note in the sequence."

The reason you have to tell the system what topic you want, of course, is because you're not connected to the topic at that particular time. If you were, the name of the topic would be included in the ACTION prompt. And remember: When you see the name of the topic as part of the ACTION prompt, you don't need to include the topic name in your command.

You also can read a range of messages by using a hyphen between the numbers. For example, if you wanted to read notes 2 through 5 in the sequence, you would type:

```
*************************************************************

    ACTION-> READ "AMERICAN HOT TUBS" 2-5 <RETURN>

*************************************************************
```

In addition, you can read a series of notes which aren't in sequential order by separating the numbers with commas:

```
*************************************************************

    ACTION-> READ "AMERICAN HOT TUBS" 1,3,6,7 <RETURN>

*************************************************************
```

Maybe you don't remember the numbers of the last three notes in the topic but you want to read the last three just the same. To do that, you'd type:

```
*************************************************************

    ACTION-> READ "AMERICAN HOT TUBS" -3 <RETURN>

*************************************************************
```

TAKING THE EXPRESS ROUTE

You can even tell an ACTION prompt to *stop* displaying ACTION prompts. Come again?

Sure. While you're reading a series of notes, seeing that ACTION prompt at the end of each one can be burdensome, time-consuming and even unnecessary. That's why a NONSTOP command is available—to present the notes you've requested without seeing the ACTION prompt after each one. (You'll still see the -More- prompt after a screen of text is displayed, however.)

Let's say you wanted to read all the notes in "AMERICAN HOT TUBS" without seeing the ACTION prompt after each note. The command would look like this:

****

ACTION-> READ NONSTOP "AMERICAN HOT TUBS" 0-8 <RETURN>

****

Note that the NONSTOP part of the command must come immediately after the READ part of the command. The same holds true if you are connected to a topic and you type READ without a topic name. The NONSTOP function will go with any of the READ commands we've shown you.

For a really fast trip through the notes, you can even eliminate the "-More-" prompt. This is particularly useful if you want to get the contents of PARTI notes or topics quickly and send them to a disk file or your printer for reading off-line. The command for READing without stopping is BATCH READ.

Suppose you wanted to read all the notes in "AMERICAN HOT TUBS" without stopping for either -More- or ACTION prompts. You would type:

****

ACTION-> BATCH READ "AMERICAN HOT TUBS" 0-8 <RETURN>

****

And if you forget the number of the last note in the sequence, that's no problem. A dollar sign ($) tells the system that you want to read through the last note in the topic. For example, if you wanted to BATCH READ all the notes in "AMERICAN HOT TUBS," you would type the following:

****

ACTION-> BATCH READ "AMERICAN HOT TUBS" 0-$ <RETURN>

****

By the way, when BATCH READing, it is good to keep in mind that CONTROL S and CONTROL Q will freeze and restart the display in PARTI, just as it will elsewhere in the system.

TURNING ON THE SCANNER

There will be times, no doubt, when you won't want to read every word in every note in a conference. That's when you'll use the SCAN command to read only a few lines of each note instead of the entire text. The format for the SCAN command is exactly like that of the READ command.

Say you want to scan a few lines of each note in "AMERICAN HOT TUBS." You would type the following at the ACTION prompt:

****

ACTION-> SCAN "AMERICAN HOT TUBS" 0-8 <RETURN>

****

You'd see the first few lines of the opener, then you'd see the ACTION on "AMERICAN HOT TUBS" 0 (of 8)-> (Next) prompt. To scan note No. 1 at this point, all you would have to do is type <RETURN>.

If at any time in the sequence you want to see the entire note during a SCAN, type CONTINUE at the ACTION prompt and the system will do just that—print the remainder of the note.

(By the way, when using the SCAN NONSTOP and BATCH SCAN commands, you won't have the opportunity to see the entire contents of the note by using the CONTINUE command, since the ACTION prompt between notes is eliminated by the NONSTOP and BATCH commands. That makes BATCH and NONSTOP more practical with READ than SCAN.)

A BACKWARD STEP

There may come a time when you find yourself reading a series of notes using the READ or SCAN command and in the middle of the sequence you decide you've read enough. That's when you can use the BACK command.

BACK is only meaningful when you are at an ACTION prompt in the middle of a discussion. It returns you to the prompt that was on your screen just before you gave the READ or SCAN prompt. It might be easiest to remember it this way: while "Next" shows you the next note in the range, BACK acts as if the note you just saw was the last one in the range and takes you back to what you were doing before you asked for the READ or SCAN.

ACTION AT THE INBOX

Perhaps you recall our discussion earlier about PARTI being a big conference hall, where we spoke about the reception area when you first enter. The area is actually called the INBOX and it's a combination on-line secretary and mail carrier. It keeps tabs on everything that concerns you in the PARTI system.

We also spoke of how you can send private mail to other PARTI users as well as read notes from topics that interest you and write notes to conferences that you are invited to join. Well, the INBOX is where all of this comes together.

In fact, the very first thing PARTI does when you enter the system is check your INBOX. After that, it gives you a PROFILE of what's in your INBOX, something like this:

****

SEGMENT	NAME	(# of NOTES)
1	Urgent Notes	(1)
2	Personal Notes	(3)
3	"AMERICAN HOT TUBS"	(8)
4	"MARTIANS I HAVE KNOWN"	(3)
5	"YUPPIES FOREVER"	(2)

ACTION on 17 Inbox Notes–> (Read)

****

Let's take time out to look at this PROFILE menu. It tells a lot about the PARTI system.

First on the list is Urgent Notes. These are private messages that other PARTIers have written to you and marked as "Urgent." The second segment of messages is also private notes written to you by other PARTI users, but they weren't marked as being urgent. (By the way, we'll talk about how you can write your own messages later in this chapter.)

Next comes a rundown on the topics you have joined (remember the "Join" command?) in which new notes have been added since your last visit. In the example above, the system is telling you that there are eight new notes in "AMERICAN HOT TUBS," three new notes in a topic you have joined called "MARTIANS I HAVE KNOWN," and two new ones in "YUPPIES FOREVER."

Note that the system suggestion at the prompt following PROFILE is "(Read)." If you agree and press <RETURN> at this point, the system would go to "Segment" number 1, present the urgent note to you and give you another ACTION prompt. If you keep pressing <RETURN> at the subsequent ACTION prompts, the system would go down the list, displaying the notes in each segment through the last waiting message in your INBOX. Then you'd return to the ACTION prompt and be told there were no more notes waiting.

Perhaps you don't want to read the segments in sequence. Maybe you just can't wait to read the latest from "MARTIANS I HAVE KNOWN." (We know we'd like to read that one first.) So, the command at the ACTION on INBOX notes prompt would be READ 4, telling the system, "Let's read all the notes in Segment 4 first." After that, you'd be taken back to the INBOX prompt where a profile would show that the "MARTIANS I HAVE KNOWN" segment has disappeared, since you have read all the notes in that topic.

The commands READ NONSTOP and BATCH READ work at the INBOX prompt as they do elsewhere. So do all the SCAN commands, just as they do if you were reading an individual topic.

In fact, it helps to think of the INBOX as forming sort of a temporary TOPIC you might call "NEW BUSINESS" where new ACTION is demanded on things you've started previously.

DELETING AT THE INBOX

There are other commands available at your INBOX. One is DELETE. You can remove notes from your INBOX, by typing DELETE. We'd suggest you use this command sparingly, though. After all, those people who send you privates notes expect you to read them quickly. If you accidentally delete private notes sent to you, there's a way to recover them and we'll talk about that later. But it's a good idea to read them instead of deleting them from your INBOX.

There may come a time when you want to DELETE INBOX notes from a topic you're no longer interested in. Let's say you have decided you're tired of the "YUPPIES FOREVER" conference. The topic is Segment 5 in your INBOX. So, at the ACTION prompt, you would type DELETE 5 and those notes would be taken out of your INBOX.

To make sure you get no further notes from the "YUPPIES FOREVER" topic, you need to LEAVE the conference. That's easy—just type LEAVE "YUPPIES FOREVER" at any ACTION prompt. As a result, no notes from the conference will be left in your INBOX again, unless you decide to reJOIN it later.

Intentional Digression Department: We've been talking about the INBOX as a stop you always make when you enter the PARTI conference hall. That's true, but it's more than that. INBOX is a command as well. No matter what your ACTION prompt looks like, no matter what topic you may be attached to, you can check on your waiting messages at any time in PARTI by typing INBOX at an ACTION prompt. In a fast-moving topic, for example, one or more messages may be added while you're on-line. This way, you can check on them at any time you're in PARTI.

THE WRITE STUFF

So far, it may not appear as if there's much in PARTICIPATE in which to participate. That's because we haven't discussed how to write notes. So, it's time to switch directions and start thinking about how to add your own two cents' worth to PARTI topics. We'll also find out how to send private messages to other PARTI users with the system's own internal electronic mail.

Knowing how to WRITE notes, of course, is as important as knowing how to READ notes if you're going to make full use of this impressive conferencing system. The good news is that using the WRITE command is easier than using the READ and SCAN commands.

The WRITE command can be issued at any ACTION prompt. Once you give the command, you're taken to an area known as the "Scratchpad." As the name suggests, this is a temporary area for your words. Think of it as a "buffer" of sorts where you compose your message before sending it on its way.

It always works the same. Whether you're writing a private message to a PARTI friend, writing an opener to a conference or subconference or adding a note to a topic, you always begin by opening the Scratchpad with WRITE.

Once you type WRITE <RETURN> you'll see this reminder:

****

Enter the text of your note, then type .send or .open to transmit.

****

In other areas of The Source, such as SourceMail, it has been necessary for you to press <RETURN> before reaching a given number of characters, usually at the end of a screen line. However, that's not necessary in the Scratchpad area. The Scratchpad contains a word-processing program that allows you to keep entering text, nonstop, until you want to end a paragraph. In fact, it even contains a "wordwrap feature" that prevents pieces of words from dangling at the end of a line on the screen. If a piece of a word is left dangling, it brings the entire word down to the next line, just as most computer word processors work.

But, if you feel better about hitting a <RETURN> at the end of a line on your screen while composing text in a Scratchpad, go ahead. The Scratchpad editor is easy to get along with—it will ignore any single <RETURN> that is typed. It only takes notice when you press two returns—<RETURN> <RETURN>. Then, and only then, it ends a line of text and starts a new line. Here's an example:

****

The Scratchpad word processor will ignore RETURNs <RETURN>
such as the one I entered on the line above<RETURN>
but if I wanted to start a new paragraph I would <RETURN>
need to get to the beginning of a new line so <RETURN>
I hit the return twice and the space bar three times.<RETURN>

<RETURN> <SPACE> <SPACE> <SPACE>
   In this way I go to the beginning of a <RETURN>
new line and indent to form a paragraph. <RETURN>

****

Remember, pressing <RETURN> just once is ignored in the Scratchpad. Press <RETURN> twice, followed by three spaces for an indented paragraph. If you want to compose in unindented paragraphs separated by a blank line, you must press <RETURN> three times—<RETURN> <RETURN> <RETURN>. Just don't get too "space happy." Folks don't go into PARTI to read blank lines.

Five basic commands can be used in the Scratchpad, though there are some lesser ones that we'll talk about later. Commands in the Scratchpad are called "dot commands" because they are preceded by a period or dot. And, as with the other dot commands you've met, they must be on a new line. In other words, they must be the first thing you type on a new line on your screen after pressing <RETURN> at least once. The basic commands are these:

- SEND Sends a note to an existing topic or a private note to another PARTI user.
- OPEN Sends the note as an "opener" for a new topic or a subtopic of an existing topic.
- ACKNOWLEDGE Sends an acknowledgment back to you when the PARTI user who is to receive the message reads it.
- URGENT Places the private note in the Urgent Notes segment of the recipient's INBOX.
- DELIVER Holds the message or opener until a date and time you specify before posting it.

SENDING A NOTE TO A TOPIC

Let's go back to the "AMERICAN HOT TUBS" topic for an example of sending a note to an existing topic. Let's say that after reading the opener and eight notes of the conference, you're ready to comment. If you were already connected to the topic, you could tell by looking at the ACTION prompt:

****

ACTION on "AMERICAN HOT TUBS" 0 (of 8)–>(Next)

****

To add your comment all you have to do is type WRITE and the system would assume you want to add what you write to the "AMERICAN HOT TUBS CONFERENCE" as soon as you finish.

If you were not at an ACTION on "AMERICAN HOT TUBS" prompt, you would type the following, remembering to put the name of the topic in quotation marks:

****

ACTION–> WRITE "AMERICAN HOT TUBS" <RETURN>

****

That tells the system that what you are about to write should be connected to the "AMERICAN HOT TUBS" topic. Once you are in the Scratchpad, write what you have to say and end it with .SEND and the system will verify where it was sent.

****

Enter the text of your note, then type .send or .open to transmit.

I've been shopping for a hot tub for several months now and am confused by all the finishes available for them. Does anyone have any opinions about what kind of finish is best? <RETURN>

.SEND <RETURN>

Sending to "AMERICAN HOT TUBS" as note 9

ACTION on "AMERICAN HOT TUBS" 0 (of 9)–> (Next)

****

Notice that you'll be taken back to the ACTION prompt you left before you entered the Scratchpad area. Notice, also, that a note has been added to the topic. That note is yours. To check it, you can type READ 9 or READ "AMERICAN HOT TUBS" 9 to see what you have written:

****

9 (of 9) YOUR PARTI NAME March 2, 22:05 Eastern (550 characters)

I've been shopping for a hot tub for several months now and am confused about all the finishes available for them. Does anyone have any opinions about what kind of finish is best?

ACTION on "AMERICAN HOT TUBS" 9 (of 9)–>(Next)

****

SENDING A NOTE TO A PARTI FRIEND

Sending a private note to a friend on PARTI is as simple. Just make sure that you're not connected to a topic when you type the WRITE command. In other words, make sure that you are at a general ACTION prompt instead of ACTION on a TOPIC prompt. That way, the computer will not assume you want to send the note to a topic.

You must know the PARTI name of the person to whom you are sending the note. This, of course, is the name the person uses on PARTI. You'll find these names in the headers of the notes and the topic openers. (There's another way to find PARTI names and we'll show you that in a minute.)

To make sure the message gets to the right person, the name must be typed exactly as it appears on the PARTI membership roster. Let's say you wanted to send a private message to Dave, whose PARTI name is DAVE PEYTON. After typing WRITE from a general ACTION prompt, here's how you would do it.

****

Enter the text of your note, then type .send or .open to transmit.

Hi Dave! I just finished reading the book that you and Charlie wrote and I liked it so much, I'm going to rush right out and buy copies for all my friends. It not only answered all the questions I had about The Source, but it cured my hiccups and made me an all-around better person. <RETURN>

.SEND DAVE PEYTON <RETURN>

****

Notice that you *don't* put the name of a PARTI user in quotes as you do PARTI topics. That's the way the system determines the difference between user names and topic names. After you enter the PARTI name of the person to whom you want to send the message and press <RETURN>, the system will acknowledge the message as being sent and give you a chance to write an optional subject line.

****

DAVE PEYTON—OK

Enter optional subject line (up to 120 characters) Your Source book <RETURN>

****

Next, the system will show you the message header as it is being sent to Dave. It will look something like this:

****

85.19876 by YOUR PARTI NAME, March 1, 1985 at 15:58 Eastern about Your Source book (224 characters) ACTION ->(Quit)

****

Hmmmmm! What's the meaning of that number—85.19876—that appears in place of the TOPIC name? Well, when you send a private message to another user or group of users, the system doesn't want a name. It merely assigns it a number. Incidentally, the first two numbers—in this case 85—are the year in which it was sent and the remainder of the numbers indicate the number of the message that was sent in that year.

When Dave finds the message listed on his INBOX the next time he drops by, he'll see this when he reads it.

****

85.19876 by YOUR PARTI NAME, March 1, 1985 at 15:58 Eastern about Your Source book (224 characters)

Hi Dave! I just finished the book that you and Charlie wrote and I liked it so much, I'm going right out and buy copies for all my friends. It not only answered all the questions I had about The Source, but it cured my hiccups and made me an all-round better person.

ACTION on 85.2413->(Next)

****

The system is asking Dave to do something with the message. Dave, being the nice person he is, probably will write a thank-you note. The system has temporarily "attached" Dave to the note you sent, just as the system temporarily attaches you to a topic. So, when Dave types WRITE at the prompt above, the Scratchpad is opened for reply to the message. After he writes the reply, he would use the .SEND command to send it back to you.

But, here's a catch! You can't reply to his reply by typing WRITE at the ACTION prompt following the reply and a simple .SEND at the end without an address. Why? Because you initiated the note originally, the system sees a reply to his reply as simply a message to yourself.

There's a way around that, of course—just end a reply to Dave's reply with .SEND DAVE PEYTON. That way, Dave would receive a new note with a new number.

SENDING TO A LIST

The great thing about the sophisticated PARTI system is that it seems to cater to your every whim. If, for example, you've decided that what you have written on your Scratchpad is a genuine work of literary art to be shared with scores, you can .SEND it to as many PARTI users as you like, even to as many TOPICS as you like, as long as you separate each recipient with a comma.

And what if all the recipients won't fit on one line? Don't worry. Just end the line with a comma, before you press <RETURN> and start typing more names and PARTI topics on the next line.

When you're finished, press <RETURN> without putting a comma in front of it.

Let's assume that you have just had a great thought and you want to share it with many. At the end of your Scratchpad, here's how you might send it:

****

.SEND DAVE PEYTON, CHARLIE BOWEN, "AMERICAN HOT TUBS", <RETURN>

"YUPPIES I HAVE KNOWN", ALLAN, SHERWIN, BREEZIE, RON,<**RETURN**> "GREAT THOUGHTS"<**RETURN**>

****

You'll be asked for an introductory line, but it will only be used in the private messages. You'll see the system verify that the message has been sent to all those places you wanted it sent. Then you'll be returned to the last ACTION prompt you were at before you had this rush of creative genius.

OPEN: THE BIG STEP

Opening a topic is a big step for most PARTI users. Whether they are opening a discussion for a few close friends or business associates or proclaiming a public conference which everyone is invited to join, it's probably the step that causes the most people to hesitate the longest.

But, after all, that's what PARTI is all about, isn't it? Electronic conferencing and discussion of topics? Then why be concerned? If everyone hesitated, PARTI would be a pretty empty meeting hall.

Before we talk about opening a public conference, we need to think a little about how the topics on PARTI on The Source are organized. We'll go into much greater detail in the next chapter. But for right now, all you have to understand is that all the public discussion topics on PARTI on The Source have sprung from one major topic appropriately called "PARTI." It is here that people come when they want to OPEN a public topic and get the most attention from PARTI users. That's because the vast majority of PARTI users are already connected to (that is, they've already JOINed) the "PARTI" topic so when you address an opener to "PARTI," all the people who have joined "PARTI" will receive the opener.

What we're saying is that almost all the major topics of discussion on PARTI are actually subtopics of the central "PARTI" conference.

Let's say, however, that you've had enough general talk about hot tubs all over America in the "AMERICAN HOT TUBS" conference and you want to start a discussion about hot tubs in the eastern United States. Plus, you want to make sure that all the PARTI users who have joined the "AMERICAN HOT TUBS" conference see the note. What you need to do is leave a note in the main conference that says, in effect, "I'd like to form a subcommittee to talk about hot tubs east of the Mississippi."

In a perfect world, the note in the main conference would also be the opener for the new subconference. Well, whether or not PARTI is perfect, it's easy and possible to make a note in the main conference be the opener in the subconference.

It begins at the Scratchpad again. After typing WRITE at the Scratchpad and getting the familiar message the Scratchpad has to offer, you might write the following:

****

It occurs to me that maybe we easterners might want to get together for our own discussion about hot tubs on the East Coast. Anyone want to join me in a discussion?

****

This time, you'd use the .OPEN command which says to send this as a special kind of note—an opener for a new conference. But you would attach it to the "AMERICAN HOT TUBS" conference so all the hot tub enthusiasts will see it. Here's how you'd do it and the prompts you'll see after you use the .OPEN command.

****

.OPEN "AMERICAN HOT TUBS" <RETURN>
Please name this topic (up to 32 characters):
EASTERN HOT TUBS <RETURN>
Enter optional subject line (up to 120 characters):
An overview of hot tubs available east of the Mississippi <RETURN>

Created as . . .

"EASTERN HOT TUBS" by YOUR PARTI NAME, March 8, 1985 at 19:17 Eastern about An overview of hot tubs available east of the Mississippi (455 characters & 0 notes)

Branching off of "AMERICAN HOT TUBS" as note 10.

ACTION–> (Quit)

****

You've done it. You've written the tenth note in the "AMERICAN HOT TUBS" conference and it is actually an invitation to conferees to join in a subconference called "EASTERN HOT TUBS." The next time everyone who has joined the "AMERICAN HOT TUBS" topic gets to note 10 in that topic, they'll see the following:

****

***** Branching off of "AMERICAN HOT TUBS" as note 10 of 10

"EASTERN HOT TUBS" by YOUR PARTI NAME, March 8, 1985 at 19:17 Eastern about An overview of hot tubs available east of the Mississippi (455 characters & 0 notes)

It occurs to me that maybe we easterners might want to get together for our own discussion about hot tubs on the East Coast. Anyone want to join me in a discussion?

ACTION on "EASTERN HOT TUBS" 0 (of 0)–> (Join)

****

The system's suggestion is to join. And if people join, they'll automatically receive future notes written to the subtopic as well as future notes written to "AMERICAN HOT TUBS." Those not interested in hearing about East Coast hot tubs would simply type NEXT instead of JOIN at the subconference prompt and they'd never receive future notes in the conference in their INBOX. That doesn't mean they couldn't read the subtopic any time they like. The command READ "EASTERN HOT TUBS" 0-$ at any ACTION prompt will put them in touch with what's happening in the conference.

OPENING A PRIVATE TOPIC

The other way to open a topic is to do it for a private group of PARTI friends who gather to discuss a topic of mutual interest. In this instance, you don't .OPEN it for all the addressees in a pre-existing topic. Instead, you .OPEN it for a group of PARTI names. Each name on the list will receive the opener to the conference and an invitation to join.

When opening a private conference, make sure you're not temporarily attached to a topic. You should be at a more general ACTION prompt when you enter the Scratchpad. After you write the opening note, the .OPEN command at the bottom might look like this.

****

.OPEN DAVE PEYTON, CHARLIE BOWEN, SHERWIN, BILLIE, FRED, <RETURN>
HARRY <RETURN>

****

You then would complete the information for the header and send the opening note on its way to those you want to invite to join.

THREE MORE SCRATCHPAD COMMANDS

There are three more commands you should know about. All three of them are used mostly when sending private messages to PARTI friends, and all of them have important functions.

.ACKNOWLEDGE To get confirmation that a note has been received by another user, type .ACKNOWLEDGE on a new line before typing .SEND or .OPEN on yet another new line. Acknowledgments appear in a special segment of your INBOX labeled "Acknowledgments." You can see a list of everyone who has read the note by typing PROFILE plus the note name or number plus the word ACKNOWLEDGEMENT at the ACTION prompt.

.URGENT To make sure that a note is given special and immediate attention by those to whom you addressed the note, type .URGENT on a new line before sending the note from the Scratchpad area. The note will then appear in the "Urgent Notes" segment of the INBOXES of those who receive the note.

.DELIVER Here's a clever one. It delays the delivery of a note until a certain date or time which you determine. Here's a way to send out your electronic Christmas cards to PARTI friends in July. Just open the Scratchpad and write your Christmas wishes any time before the holidays. Then, before you .SEND the contents of the Scratchpad to your list of friends type the following on a new line:

****

.DELIVER 12/25/86 08:00 <RETURN>

****

The note won't go in the recipients' INBOXes until Christmas morning, 1986.

By the way, this also is a good way to write reminders to yourself about important events. Simply enter the Scratchpad area and write a letter to yourself about an upcoming birthday or anniversary. Then command the note to be delivered a day or so before the event and .SEND it to your own PARTI name. The note will appear in your INBOX on the date you specify.

A LITTLE PEP TALK

Is your mind reeling yet? How in the world can you memorize all these commands?

Well, you can't. But then again, you don't have to. Even PARTI's old hands don't know *all* the commands. What makes PARTI so powerful is its sophisticated, extensive command system. It has commands for all occasions—and on *some* occasions you might have to just look them up. Also, there's extensive on-line help available. Stick around, because we'll be telling you how to get to it.

Meanwhile, here's some good news, friend: The commands you've learned in the first two-thirds of this long chapter will see you through perhaps 90 percent of the day-to-day reading and writing of PARTI. What you'll see from here through the end of the chapter are additional commands that can make PARTI life easier, more enjoyable, more meaningful.

In other words, if this were a highway, you'd likely see a sign about now that said, "Begin Browsing." With some of these commands, it's enough to know where to find a discussion if you need them; others aren't essential, but are useful to be aware of. For instance. . . .

EDITING IN THE SCRATCHPAD

As with most word processors, PARTI's Scratchpad has some editing features available that help you be more efficient at writing clear and concise notes and messages on the system.

As with all the commands we have explained so far, the editing commands begin with a dot (or period) and must be placed on a new line. If you'll recall, the dot must be the first character typed on the new line. Here are some more of the dot commands:

.CLEAR This command erases the entire contents of your Scratchpad and allows you to begin all over.

.EDIT This command takes you to The Source editor that we discussed in another chapter.

.DISPLAY If you want to see what you've written in your Scratchpad so far, type .DISPLAY. This is a good way to see how your note is going to look before sending it. After the contents of the Scratchpad are displayed, you can continue typing text, or use any other dot command.

.CENTER If you want to center the text on the page as a title or to emphasize it, you can type .CENTER followed by a space and the text. By hitting <RETURN> once after you type the text you want centered, you'll return to text which is flush left with the screen. This is the only instance, by the way, where the PARTI Scratchpad recognizes a single <RETURN>.

.READ You can actually read notes from within the Scratchpad with this command. All you need do is type .READ plus the message number or topic name followed by the note number or range of note numbers. After the note or notes are displayed, you can continue typing text. The text of the note or notes you read will not be displayed in the note you are writing.

.COPY You can copy any note in the system you have access to and put it in your own note by using the .COPY command. For example, if you wanted to copy the third note in "AMERICAN HOT TUBS" and make it part of your note, you would type:

****

.COPY "AMERICAN HOT TUBS" 3 <RETURN>

****

on a new line of the Scratchpad and it will appear in the text of the note you are writing. You can only copy one note at a time. To copy another note, you must use the .COPY command again on a new line followed by the note you wish to copy.

.COPY FILE, followed by a file name from your Source file area, allows you to copy any file in your own storage area and put it in the note.

.FILE, followed by a name, allows you to copy what you have written in the Scratchpad and put it in your own file storage area on The Source.

THE PARTI PROFILE

The PROFILE command gives you the ability to "look behind the scenes" to get descriptive information about how you are known on PARTI, other users on PARTI and about the various topics being discussed on the system.

One of the most useful combinations uses the command PROFILE MYSELF. It gives you a lot of information about how PARTI sees you as well as a history of your activity on PARTI. When you type PROFILE MYSELF at the ACTION prompt, you'll see a menu which includes:

****

1. Terminal information
2. Accounting information
3. List topics you have joined
4. List topics you have initiated
5. List messages you have written
6. List files in your current directory

****

You can select one of the options by typing the appropriate number. If you ask for:

• Terminal information, the system will report the number of lines you scan when you use the SCAN command, the number of lines printed before you see the -More- prompt (called "line pause") and the screen width in which the notes are printed.

• Accounting information, you'll be told the total number of hours you have been in PARTI and the hours you have used PARTI since the last bill was issued.

• The topics you have joined, you'll see a list of the conferences you have JOINed to receive notes in your INBOX.

• A list of topics you have initiated, you'll see a list of the topic names, the date the topic was initiated and how many notes each topic contains.

• A list of the messages you have written, you'll see a list of the private numbered messages you have written, when they were written, and the subject line.

PROFILING OTHER PARTI USERS

You can get a profile of any other PARTI user by typing PROFILE USER and the PARTI user name of the person at the ACTION prompt.

How much other users tell you about themselves is entirely up to them. Some gladly list their names, addresses and phone numbers and other biographical information using a MODIFY command, which we'll investigate shortly. Others only list part of that information. Some list none. It's entirely up to the individual user.

You also can profile several users at once. Just type PROFILE USER and the user names separated by commas.

To see which users currently are on-line in the PARTI system, type PROFILE ONLINE, which is similar to the ONLINE command at The Source command prompt.

A PARTI FULL OF SWITCHES

Now we don't mean to get kinky here, but you ought to know that PARTI users have "switches" at their disposal.

These switches, however, are the kind that turn PARTI functions off and on.

For example, the new user is introduced to PARTI through a menu. That menu can be turned on and off. In addition, you've seen how the new user always sees a suggested command in parentheses following the ACTION prompt. That switch can be turned off when you're tired of having the system get suggestive.

As a new user, you will always see the PROFILE menu of the notes in your INBOX each time you enter. That's called an AUTOPROFILE. It can be turned off if you ever get tired of seeing that menu.

When you're in the Scratchpad, most new users see a "confirmation" before they send a note to a topic. In other words, the system says "Are you sure you want to send this to that topic?" You must answer Yes or No. This confirmation switch can be turned off.

When you use the command PROFILE switches, you'll see the "position" of all these switches mentioned above plus whether they are "off" or on," in addition to how many lines you scan when you use the SCAN command, how many lines are displayed before the -More- prompt and the width of your terminal screen.

FINDING OUT WHO'S HERE

You can find out the PARTI names of everyone who has joined a particular PARTI topic by typing PROFILE plus the topic name (in parentheses) plus the word MEMBERS at the ACTION prompt. You'll see a list of PARTI names of people who have joined the topic and the note in the conference they last read in parentheses.

If you type PROFILE plus the topic name plus the word ADDRESSES at the ACTION prompt, you'll see a list of all addresses to which the topic was sent. Most of the time, public topics are "addressed" to another topic, so the address will be a topic name.

RETAINING A TOPIC

The standard length of time a topic is kept on-line before it automatically disappears is set up by the administrator on the PARTI system, but it can be modified by the originator of the topic, as we'll see later. To see the retention period for any topic type PROFILE plus the topic name plus RETENTION.

There may be times you'd like to see how many subtopics have started from a topic. If you type PROFILE plus the topic name plus BRANCHES you'll see all the branches off that main topic and the number of notes in each of the branches.

If you want to see if you have access rights to a particular topic, type PROFILE plus the topic name plus ACCESS.

If you care to check on the list of recipients who have read a note you sent them after using the .ACKNOWLEDGE command, type PROFILE plus the note number plus the word ACKNOWLEDGMENTS. You'll be given the names of those who received it and the date and time the note was read.

MODIFYING PARTI TO SUIT YOUR NEEDS

In the last section we talked about turning switches off and on to suit your needs on the system. There are more than just switches to turn off and on. In fact, there is a long series of MODIFY commands that let you fine tune the system. For instance:

MODIFY PROFILE This is one of the most used MODIFY commands. When you type it at the ACTION prompt, you'll be shown this menu:

****

Which part of your profile do you want to modify?

1. Address
2. Phone
3. Description

****

If you choose option 1, you'll be prompted for your mailing address. Choosing option 2 will prompt you for your phone number and option 3 is where you can write a description of yourself. This information is provided to other users when they use the PROFILE USER command we discussed in the last section.

MODIFY MENU This is one of those switches we talked about. If you currently receive a menu when you enter PARTI and you want to turn it off, type MODIFY MENU and it will disappear. Type MODIFY MENU again and it will return.

MODIFY BACKSPACING This switch is the one that removes the parenthetical suggestions at the ACTION prompt.

MODIFY AUTOPROFILE This is the switch that turns the INBOX profile off and on.

MODIFY CONFIRMATION This is the switch that turns the confirmation line off and on in the Scratchpad.

MODIFY SCANLINES This one changes the default for how many lines you scan when you use the SCAN command. You'll be asked to enter a new scan line count at a prompt when you enter this command.

MODIFY LINEPAUSE This one lets you change the number of lines displayed on your terminal before seeing the -More- prompt. If you set the LINEPAUSE at 0, there will be no -More- prompts ever.

MODIFY WIDTH You can change the maximum number of characters you want on any line of output using this command. You'll be prompted for the new line length when using this command.

MODIFY "TOPIC" ADD With this command, you can send a topic opener you have created to an additional list of addressees. You simply type the topic name where you see "TOPIC" above and add the addressees after the command. The addressees must be separated by commas. Remember that this doesn't JOIN the additional addressees to the topic—it just invites them to join. Remember, too, that an addressee can be a topic as well as another user name. If you address it to a topic, it becomes a subtopic of that main topic.

MODIFY "TOPIC" REMOVE If you want to remove addresses from a distribution list of addressees who are supposed to receive an invitation to join, use this command followed by the PARTI names of those you want to remove.

MODIFY "TOPIC" UNJOIN This command, followed by a user name or list of names, removes users from any topic you have created.

MODIFY "TOPIC" RETENTION This command allows you to alter the time a topic you have created remains in the PARTI database. When you use this command, you'll be asked two questions. First you will be asked how many days a topic should be retained after it was last referenced by a user. Each time someone reads a topic note or sends a note to it, that counts as a reference. The next question is for how many days after it was started should the topic be retained. A topic will be retained in the PARTI database as long as it satisfies both retention periods.

MODIFY "TOPIC" OPENER You can change the wording in the opener of a topic you started with this command. When you type this command, you're taken to the Scratchpad where you type the new opener and use .SEND to replace the old opener with the new one you've just written.

MODIFY "TOPIC" SUBJECT This command allows you to change the optional subject line of a topic you have created.

MODIFYING ACCESS TO YOUR TOPIC

You're probably getting the message by now that PARTI is a powerful system with the commands necessary for you to be able to tailor it to suit your needs. PARTI is rightfully proud of this flexibility, we think.

Those powers are never more evident than in dealing with the topics that you initiate on the system. You've already seen how you can invite as few or as many people as you like to join the topic, how to add and delete people from the conference, how to write the opener, change it later if you like and set the period of time it is held in the database.

But as a "topic czar," you have even more powers to control topics you have started.

By using the command MODIFY "TOPIC" ACCESS (and inserting your topic name in place of "TOPIC," of course), you enter a menu area of the system where you can further modify your topic to place more rules on what topic JOINers can do and not do in your topic.

The first menu in the MODIFY "TOPIC" ACCESS area gives three options. The first two options allow you to add or delete one specific topic addressee. The second option allows the same functions for a number of addressees.

The third option, labeled "Non-addressees" is the most important, however. By choosing it, you can do the following:

- Make it a "read only" topic, which means that those people you invite into the topic are the only ones who can read the messages you write. They can't add their own notes to the topic.
- Allow those you have invited into the topic area to read the notes and create branches to the main topic but not send notes to the main topic.
- Allow JOINers to write notes but not read other people's notes.
- Customize through a third-level menu to deal with such functions as what the JOINers can and can't read in the topic, whether or not they can profile members, etc.

The standard (default) for PARTI on The Source is that all topics allow JOINers to read everything, write notes, create branches and have all the major commands we've mentioned in connection with topics at their disposal. And that's the way it should be for most topics. But it's important to know that if you ever want to place restrictions on a topic, you can by using the MODIFY "TOPIC" ACCESS command.

AND THAT'S NOT ALL . . .

Wait! Before you toss the book in the air and decide to take up knitting as a hobby, keep in mind what we've said. The fact is that, as the average PARTI user, you'll not need to know more than a third of these commands.

If you have understood the concepts we have presented, you've learned more important information than if you had memorized all the commands.

We've talked to many PARTI users—neophytes as well as experienced users—and they all tell us the same thing: Once you understand the *structure* of PARTI, the commands are easy to master. And there's so much on-line help, finding the specific way to write the commands is easy.

But to be sure you have these major commands on hand the first few times you enter PARTI on The Source, we'll include them in a list of commands at the end of this chapter.

There are still a few more commands we'll be discussing in the next chapter. You'll find them in our list at the end of this chapter. So if you don't recall seeing a discussion of them in this chapter while browsing the list, have patience. We'll get around to discussing them in the next chapter, when we'll take a much closer look at PARTI as it is practiced on The Source.

MOST COMMONLY USED COMMANDS

? gets a list of available options.

HELP gets a list of options with a brief explanation for each.

?(COMMAND) gets a detailed explanation of a specified command or concept.

READ NONSTOP reads through notes without pausing after each note.

READ "TOPIC" reads the named topic opener.

READ "TOPIC" *note #* reads the specified note number in topic.

READ *segment #* reads the specified segment of your INBOX or Found Notes.

READ FILE *filename* reads the specified file from your Source file area.

SCAN (# lines) scans the specified number of line without pausing; usually reads only header and first few lines of a topic.

SCAN "TOPIC" *note #* scans a specified note.

SCAN *segment* scans a specified segment of your INBOX or Found Notes.

SCAN FILE *filename* scans a specified file in your Source file area.

NEXT reads or scans your next note.

JOIN "TOPIC" joins a topic to receive future notes.

LEAVE "TOPIC" stops receiving new notes from a topic in your INBOX.

WRITE enters the Scratchpad.

PROFILE finds information about users, topics or segments of INBOX or Found Notes (see separate section on PROFILE).

MODIFY changes characteristics about yourself or topics you created (see separate section on MODIFY).

QUIT quits PARTICIPATE.

OTHER COMMANDS

NOW sends a real-time message to a PARTI user on-line at the same time you are. PARTI name should follow NOW, followed by <RETURN>, a one-line message and <RETURN>.

ACTION returns to the ACTION prompt.

BATCH READ or BATCH SCAN reads or scans without pausing for action or for the -More- prompt.

CONTINUE reads the remainder of a note after scanning part of a note.

BACK takes you back to what you were doing before.

INBOX gets an update of your INBOX notes.

FIND SINCE *date (time)* finds notes sent to you since a specified date (and time).

FIND BETWEEN *date (time)* AND *date (time)* finds notes sent to you between specified dates (and times).

FIND (FROM *username*) (TO *username*) finds notes you received from or sent to a specified user.

FIND */string/* IN "TOPIC" (*note*) finds a specified string.

FIND */string/* (BRANCHES # *depth*) finds a string in the specified number of branching levels.

TIME gives the current date and time.

KEEP "TOPIC" note saves a note in your INBOX for future reference.

FILE *filename* "TOPIC" (*note*) stores notes in a file.

FILE *filename* PROFILE *command* stores the results of a PROFILE command in a file.

DELETE "TOPIC" deletes an entire specified topic.

DELETE "TOPIC" *note* deletes a specified note.

DELETE FILE *filename* deletes a specified file.

DELETE *segment* deletes a specified segment from your INBOX or Found Notes.

VOTE "TOPIC" votes on questions in a topic that is a poll.

QQUIT quickly exits PARTICIPATE without checking INBOX notes.

BYE exits PARTICIPATE and logs you off The Source.

SCRATCHPAD COMMANDS

Sending notes

.SEND sends a note to a topic or user.

.OPEN starts a new topic.

Options to be used before .SEND and .OPEN

.ACKNOWLEDGE obtains indication that a note has been received.

.URGENT marks a note to appear in the special Urgent INBOX segment.

.DELIVER mm/dd/yy (hh:mm) delays delivery until specified date (and time, which is optional).

Obtaining and storing notes from Scratchpad

.READ "TOPIC" *note* shows the contents of a specified note.
.COPY "TOPIC" *note* inserts text from a specified note into Scratchpad.
.COPY FILE *filename* inserts text from a specified file into Scratchpad.
.FILE *filename* stores the Scratchpad contents in a file.

Editing notes

.CLEAR erases Scratchpad contents.
.DISPLAY displays Scratchpad contents.
.CENTER centers text.
.EDIT calls up The Source editor.

Other Scratchpad commands

? gets a list of available Scratchpad commands.
.HELP gives an explanation of commands used in the Scratchpad.
?.*command* gives a detailed explanation of the specified Scratchpad command.
.PROFILE finds information about users and topics (as with the PROFILE command).
.MODIFY changes characteristics about yourself or topics you created (as with the MODIFY command).
.QUIT leaves the Scratchpad with an option to save contents.

PROFILE COMMANDS

PROFILE finds information about users, topics or segments of the INBOX or Found Notes.
PROFILE MYSELF finds information about terminal settings, accounting data, topics, messages and files.
PROFILE USER *username* shows the personal profile of a specified user.
PROFILE ONLINE lists users who currently are on-line.
PROFILE SWITCHES shows current status of your switches.
PROFILE "TOPIC" gives information about topics.
PROFILE "TOPIC" MEMBERS lists members of a specified topic.

PROFILE "TOPIC" ADDRESSEES lists addressees of a topic.

PROFILE "TOPIC" RETENTION shows retention periods associated with a specified topic.

PROFILE "TOPIC" BRANCHES shows a list of all branches of a specified topic.

PROFILE "TOPIC" BRANCHES (# *depth*) shows the specified number of branching levels.

PROFILE "TOPIC" BRANCHES (# *start note*) shows the specified number of branching levels starting with note indicated.

PROFILE ACCESS tells what access rights you have to a topic.

PROFILE ACCESS *username* shows access rights of a specified user to a specified topic.

PROFILE ACKNOWLEDGEMENTS gives a list of users who have read notes marked for acknowledgment.

PROFILE VOTES shows voting results of a poll.

MODIFY COMMANDS

MODIFY changes characteristics about yourself or topics you created.

MODIFY PROFILE enters or changes your self-descriptive information.

MODIFY PASSWORD changes your current password.

MODIFY SWITCHES controls your interaction with PARTICIPATE.

MODIFY "TOPIC" changes characteristics of a specified topic.

MODIFY "TOPIC" ADD *addressee* list sends topic to additional addressee(s).

MODIFY "TOPIC" REMOVE *addressee* list removes addressee(s) from a topic's distribution list.

MODIFY "TOPIC" UNJOIN *member* causes member(s) to leave a topic.

MODIFY "TOPIC" RETENTION changes retention period you have specified for a topic.

MODIFY "TOPIC" ACCESS changes access rights associated with a topic.

MODIFY "TOPIC" OPENER changes the text of a message or topic opener.

MODIFY "TOPIC" SUBJECT changes the subject line of a topic.

MODIFY "TOPIC" LIMIT sets limits on the number of characters allowed in any note sent to specified topic.

CHAPTER 11

PARTI on The Source

This may be hard to believe, but the PARTI seminar in Chapter 10 was pretty general, despite the fact that we gave you reams of information about commands. It was general because we spent the entire time discussing the PARTI structure—a structure you might find on other computer systems besides The Source.

Wait a minute. Does that mean PARTI isn't exclusive to The Source? That's right. The program called "Participate" is owned by Participation Systems, Inc., and is a general electronic conferencing system for several types of mainframe computers and minicomputers. Many corporations, big and small, use Participate as their major in-house electronic communications system. It offers both a private electronic mail and multiuser conferencing.

That means you may run into PARTI on systems other than The Source. If you do, you're now way ahead of the game, because you've already learned the major concepts and commands.

And because those commands have been tucked away in these two chapters, you should never, never, give this book to a friend. Keep it for your very own—let your friends buy their own copies. (We learned that concept in a PARTI topic called "MARKETING STRATEGY" . . .)

In this chapter, we'll be getting specific about PARTI on The Source—how Source subscribers have used PARTI in the past and how they use it these days.

In addition, we're going to tell you as much as we can about what you're likely to see in your early on-line experiences with PARTI on The

Source. Notice that we say we're going to tell you what you're *likely* to find. The changing nature of PARTI prevents us from telling you exactly what topics you'll run into, what the hottest topics on the system will be when you first enter the PARTI area and where you should go first to find what you're looking for.

Believe us, if there were a way to freeze PARTI on The Source just for you, so you could see the things on the screen exactly as we write them in the book, we'd do that. But, as we've observed more than once, the very nature of The Source demands that nothing remains constant. The immediacy of the medium probably is one of the things that attracted you in the first place.

So it is with PARTI, perhaps more so than other areas of The Source. We've seen times when a single news event has sparked a half dozen new topics in an evening. We've seen times when a public topic has been so hot one week that a hundred notes or more have been added in a few days. The following week, everyone loses interest and the conference dies a lonesome death unless PARTI officials decide to preserve it in the on-line archives.

Fortunately, there are some enduring fixtures in the PARTI conference hall. We'll take a look at them and tell you how you can check into them frequently to find out what's happening.

And, finally, we'll give you a few more commands you'll find helpful as you become an experienced PARTIer.

A BIG TRANSITION

And speaking of changes. . . .

For us, it's certainly satisfying to put the finishing touches on a book like this and know that once we've fine-tuned those final few words and sentences, we can clear our minds for a new project. Well, when we thought we'd done just that with this book, the unexpected happened. Or, as we say in Appalachia, "Up jumped the skunk!"

We found out that about the time this book was to be published, The Source and Participation Systems, Inc., were planning to install a completely new version of PARTI software. And we mean "completely"—the PARTI concept remained the same, but there were massive changes in the PARTI commands and menu structure between the old version (known as 3.4) and the new, called Version 4.

Well, it was back to the drawing board—er, keyboard—for a quick rewrite of these chapters to reflect the new version.

What does all this mean to you? Several things:

If you are one of the early purchasers of this book, there's a possibility you may log into PARTI and see that Version 3.4 is still there. If that's the case, have patience. It won't be long before the new version—the one we talk about here—will go into effect, and with these two chapters (10 and 11) you'll be all set.

In a few cases, the options you read about here may not be exactly as you see them on PARTI on The Source. How come? Version 4 offers the "system administrator" several options for tailoring PARTI. In this case, the system administrator is The Source. Discussions about the particulars of some of these options were continuing as we were writing.

But there's good news. There's lots of on-line help for Version 4. See the command list in Chapter 10 to see how to get help.

Even better news, perhaps, is that there are people on PARTI who stand ready to answer any question you have about the system and you can write them a private message and get a response—often within a few hours. We'll show you how.

ONE BIG HAPPY CONFERENCE

When Dave's kid was 4 years old, he was fascinated by the thought there was only one road in America.

Dave Jr. figured out that all the nation's roads were connected. They might be called by different names, might be covered by different materials—concrete, asphalt, rocks and gravel—but they were all extensions of a single interconnected highway system. (The thought kept him amused for the better part of an hour on a cross-country trip—a record which still stands, incidentally, although he's now 14.)

Well, the topics on PARTI are numerous and varied, just like the highways in America. But, in fact, PARTI is just one big happy topic with scores of subtopics, subsubtopics, and beyond. And it's a system that keeps growing and changing with every new note that's added.

You're already familiar with interconnected systems. After all, that's what you've found The Source's menu structure to be. There are hundreds of menu options, but they are all interconnected, part of a single branching menu that begins at the top of the system when you type "MENU" at the -> prompt.

The top of PARTI—at least as far as the public topics go—is a topic called "PARTI." It is the conference opener most used to open new topics. The reason is simple. Almost all PARTI users are "connected" to this topic. That means they have JOINed it, by typing JOIN "PARTI" at an ACTION prompt. Thus, when anyone writes an opener in the Scratchpad and types .OPEN "PARTI" as a dot command at the bottom of the Scratchpad, that topic opener goes to practically everyone on the system, along with an opportunity to join the conference.

This is perhaps the most important concept to learn about the content of PARTI on The Source. Once you understand the connection between all of the public topics on PARTI, the rest is all downhill.

HOW TO GET THERE

Well, we think you're nearly ready to enter PARTI on The Source. This won't be one of our on-line tours, primarily because of the changing nature of PARTI. But we'll walk you through what are likely to be your early experiences on PARTI.

Entering PARTI is as simple as typing PARTI <RETURN> from the command prompt on The Source. But before you do that, you ought to give some thought to what you want your PARTI name to be because there's a good chance the system will ask you to give yourself a PARTI name immediately. If it doesn't, it means that your PARTI name will be your Source User ID until you change it.

What do most folks use for PARTI names? There's really no standard, except that it must not be longer than 32 characters and it can't have any punctuation in it. Only a few use their complete names as their PARTI names. Many use their first name and the initial of their last name. Others simply use their first name. Others use descriptive "handles."

Whatever you decide to use, have second and third choices in mind because if the name you choose already is in use by another PARTIer, you'll have to make another choice. Remember, the shorter the name, the easier it is to remember. Also remember that you can modify your PARTI name later, if you like, though it's better if you continue to use the name you start with on PARTI.

Also, remember that if you want to give more information about yourself, you can use the MODIFY command to write descriptive information about yourself—as little or as much as you like.

Under the old version (3.4), most new PARTIers were also asked to enter such information as the width and depth of their terminal screens and to register descriptive information about themselves the first time they logged on. We've been told that the "questionnaire" is to be removed from Version 4. If, however, you see it, we recommend you go ahead and set the screen width and depth (or "linepause"), but simply enter <RETURN> when the system asks for descriptive information. You'll have time later to register that information with the system after you get more comfortable with it.

AN INBOX FULL OF NOTES

Once you get past the main PARTI gate, you're likely to see a menu. This menu contains text files that give an overview of PARTI, some instructions and some examples of some of the topics discussed on PARTI.

Read them, if you like, along with any of the other topics you see. When you finish, you'll be returned to this main PARTI menu.

But you're not there primarily to read text files from the system. What you want to do is "Begin to Participate" in "PARTI," so you should choose that menu option. This takes you directly to the conference hall reception area where you may see a surprise—an INBOX already full of notes.

That's a bit disconcerting to some new PARTI users. After all, how did the system know you were going to log into PARTI? Is there someone out there saying, "Here comes another one. Let's put some mail in his INBOX!"

Not at all. All new users get some instructive and beginning notes in their mailboxes to help them get organized quickly.

As you read these notes, here are some of the things you are likely to see.

YOUR WELCOME WAGON

One of the topics that's likely to be in your INBOX when you first enter is "WELCOME WAGON." As the name suggests, it is for all new users and it provides some important instructions as well as a way to practice using the PARTI system in the early stages.

While in "WELCOME WAGON," you'll get the opportunity to learn the fine points about getting started on PARTI. In addition, it provides an area where you can write a note to your "WELCOME WAGON" host, using the Scratchpad and the .SEND command.

PRACTICE

We can understand that a lot of people like to practice the commands they've read about before they feel completely comfortable with a computer program. If you're one of those folks, then the "PRACTICE" conference is for you.

You don't need to JOIN "PRACTICE." All you need to do is type READ PRACTICE at any ACTION prompt and the system will take you by the hand from that point. Each note in the "PRACTICE" topic explains a PARTI concept, displays it to you and tells you exactly what to do when you get to the end of the note you are reading. This may be helpful to you since it demonstrates vividly all we've been talking about so far.

JOINING THE "PARTI"

The third note that's usually in the new user's INBOX is the "PARTI" note. This is the one you need to take action on immediately, since JOINing it will put you in the mainstream of the new public topics which are being added to the system daily.

If "PARTI" pops up in your INBOX and you READ the opener to it, you'll see the ACTION on "PARTI" -> prompt at the end of it. The system's suggestion, in parentheses, will probably be "Join." You should type <RETURN> to JOIN the topic. You could do the same thing from any ACTION prompt by typing JOIN "PARTI."

Remember though, you won't get any *previous* topic openers that have been connected to "PARTI" in your INBOX. You'll only get the *new* ones which people connect to "PARTI" from that time on. And it's a good thing, since there have literally been thousands of topic openers connected to "PARTI" since the PARTI system was installed at The Source in 1982.

A lot of those early topic openers have long since disappeared. But as they were purged from the system for lack of action in them, new ones were introduced.

"PARTI" is a specially constructed topic. You can't write ordinary notes to it. You can only write branch openers to it. In other words, if you compose a note in your Scratchpad and type .SEND "PARTI" at the end, you'll be told you can't do that. Why? Because PARTI administrators don't want the important "PARTI" topic littered with notes, just topic openers.

There's a place to send notes to the general PARTI population, however. It's a conference called "PARTI BOARD," and most of the people who are connected to "PARTI" have elected to join "PARTI BOARD" as well. This topic is what the name implies. It's a bulletin board for PARTIers. Quite often you'll see messages telling you about things happening on public topics that you may have failed to JOIN.

MORE IMPORTANT TOPICS

The beauty of PARTI on The Source is that you can get as involved as you like. The more of the following conferences you join, the more involved you will become. JOINing these conferences isn't absolutely necessary, but, of course, joining them *will* help keep you in touch with what's going on in the hall since new notes added to the conferences will be delivered to your INBOX.

"PSI BULLETINS" You probably automatically joined to this topic when you first entered PARTI on The Source. To make sure, you might want to type JOIN "PSI BULLETINS" at an ACTION prompt. This is a "read only" conference, which means you can't write notes to it, nor can you start branches from it. "PSI BULLETINS" is the place where Participation Systems, Inc. places announcements about the PARTI system. News of changes to the system are listed in the notes here as are any problems that the system may have experienced since you last logged on. It's a handy reference guide to what's happening with the PARTI structure.

"NEW USER FORUM" This is a branch of "WELCOME WAGON"

and it's a place for new users to get tips on using the system and for asking for help. It's only one of the many places on PARTI where personal help is available.

"CHIMO" This is an Eskimo word that means "I am your friend." (Honest.) This topic is a weekly on-line newsletter for PARTIers about what's happening on The Source. Included is a list of most active topics for the past week, topics that have been added and the old topics that have been purged from the system during the same time. If you JOIN "CHIMO," the opener for each weekly newsletter will be added to your INBOX automatically each week.

"INDEX" Here's one of the most useful topics for finding the topics that were started before you joined PARTI and which ones are still going strong. It's called "INDEX" and by reading all of its notes and branches, you'll get an overview of all the active topics according to categories. If you JOIN "INDEX" you'll automatically get updates of topics that have been added to or deleted from "INDEX."

A SPECIAL TOPIC FOR OUR READERS

As a reader of this book, you're invited to an electronic "MEET THE AUTHORS" party on PARTI. Dave has volunteered to open a public topic with this name and attach it to the public "PARTI" topic. That means that all you need do to see the topic is type READ "MEET THE AUTHORS" at the ACTION prompt. We hope you'll join the topic and .WRITE a note telling us what you think of the book. We'll drop by often and we hope to see your note left in our INBOXes.

WHERE TO FIND THE LIFEJACKETS

Despite all our best efforts, you still may find something that stumps you while using PARTI. Well, help is as close as a note to HELPER.

So who's HELPER? HELPER is a sort of generic name for a group of volunteer PARTI enthusiasts who roam the system on The Source looking for people to help. We're going to show you exactly how to write a note to HELPER because it's one of the easiest and fastest ways of getting on-line help. You'll get a reply almost always within 24 hours and sometimes within an hour or two. And it's help from a real live person. What more can you ask for from a computer?

If you read the last chapter carefully, you probably have a good idea how to write a message to HELPER. But here's a step-by-step guide:

1. From a general ACTION prompt—not the ACTION prompt that shows you're connected to a topic—type WRITE <RETURN>. That will take you to the Scratchpad.

2. Once in the Scratchpad, type in your question.

3. When finished, use <RETURN> to go to a new line and type .SEND HELPER <RETURN>. (Since Helper is a person, that is, a PARTI name, and not a topic, don't put quotation marks around it.)

Your message will be posted in the HELPER INBOX for attention the next time a HELPER logs on. The HELPER's subsequent answer to you will appear in your INBOX under "Personal Notes."

WHEN THE PARTI'S OVER

When the time comes for you to leave PARTI, you have three choices. If you type QUIT at an ACTION prompt, the system will check your INBOX one more time. If there are unread notes there, you'll be told before you leave, to give you a chance to read them. If there are no unread notes in your INBOX, you'll be taken out of PARTI to The Source command prompt.

If you type QQUIT, the system will not check to see if there are unread notes in your INBOX. You'll be taken back immediately to The Source command prompt.

If you want to leave the system entirely from within PARTI, type OFF at the ACTION prompt and, after checking your INBOX, you'll be disconnected from The Source.

MORE COMMANDS FOR YOU

Just when you thought you'd read about all the commands on PARTI, here we come with even more for you to remember.

That's not really the case. You don't have to remember *any* of these, in fact. They're handy only in special situations. They're the kind you should be aware of. Then, if you need one of them someday, you'll know where to look them up.

FINDING LOST NOTES

Once you process the notes in your INBOX, they disappear from that important reception area, but they're not gone forever. They're still in the database. Retrieving them is made simple with some important FIND commands at your disposal.

The FIND command can be used to search for notes that appeared in your INBOX at some date and time in the past. Let's say it's June 25, 1986, and you want to find a private message that was sent to you by a friend the day before. You know it was sent after 6 P.M. on that date. So, at the ACTION prompt, you would type:

****

FIND SINCE 06/24/86 18:00 <RETURN>

****

That tells the system to retrieve all the notes written to you since June 24, 1986, at 6 p.m. (Note the date has to be written in the format you see with the time in military time.) If you don't specify a date, PARTI will assume today's date. If you don't specify the time, PARTI will assume you mean 00:00 hours, that is, the beginning of the day.

You can also retrieve notes written BETWEEN certain dates using the FIND BETWEEN command. For example, if you wanted to retrieve all the notes sent to your INBOX between June 1 and June 3, you would type:

****

FIND BETWEEN 06/01/86 AND 06/03/86 <RETURN>

****

You can use FROM and TO with the FIND option to FIND a specific note you sent TO a user or received FROM another use. For example:

****

FIND SINCE 06/02/86 13:00 FROM CHARLIE <RETURN>

****

****

FIND BETWEEN 06/02/86 AND 06/15/86 TO CHARLE <RETURN>

****

Not only can you find notes, but you can use a form of the FIND command to search notes in a topic for words or phrases up to 58 characters.

Let's say you wanted to search the "AMERICAN HOT TUBS" topic for all notes that mention redwood decks. From the ACTION prompt, you would type:

****

FIND/redwood decks/ IN "AMERICAN HOT TUBS" 0-$ <RETURN>

****

The system will search every note in the topic and find the ones mentioning redwood decks.

So, what happens to found notes? Well, they're placed in your INBOX in a special segment called "Found Notes." If you ask for a search and it is successful, the subsequent ACTION prompt will display:

****

ACTION on Found Notes–>

****

You can do other things before reading them if you like. When you want to deal with them, type INBOX and you'll see them in the Found Notes section.

DELETING NOTES

There may come a time when you send a note to a topic and decide later to withdraw it. Or, if you have started a topic, you may decide it's best to delete a note sent by another user to your topic. To do either, you use the DELETE COMMAND.

In both instances, you simply type DELETE, the topic name in quotation marks and the note number. Remember, you can't delete notes from others if you didn't start the topic. In someone else's topic, you can delete only your own notes.

KEEPING NOTES

Perhaps you've run across a note you want to file temporarily so you can deal with it later. The command for doing that is KEEP. For example, if you want to keep note 4 from "AMERICAN HOT TUBS" to reread later, type the following at the ACTION prompt.

****

KEEP "AMERICAN HOT TUBS" 4 <RETURN>

****

The next time you go to your INBOX, it will be there under a section called "Kept Notes."

You also can keep a note and specify when you would like it returned to your INBOX. For example:

****

KEEP "AMERICAN HOT TUBS" 4 FOR 06/18/86 <RETURN>

****

That would deliver the note to your INBOX on June 6, 1986.

THE NOW COMMAND

The NOW command allows you to send a real-time, one-line message to a fellow PARTIer who is on-line the same time you are. To find out who is currently on-line you type PROFILE ONLINE.

Once you have established whether the person you are trying to reach is on-line, you type NOW followed by the user's PARTI name and press <RETURN>.

On the next line, type the brief message and press <RETURN>. Under normal circumstances, the other user would see the message immediately.

If you don't want to be interrupted with such messages, all you have to do is type MODIFY SWITCH NOW, to turn the NOW switch off. To turn it back on again, type MODIFY SWITCH NOW again.

MODIFYING YOUR ACCOUNT: A BIG STEP

When you first enter PARTI, you enter as a "single-user account." That means you can have only one PARTI name assigned to your Source User ID.

However, many of your fellow PARTIers have several names using the same Source User ID. There are several reasons for this. A business office might share a single Source account but several people in that business may want separate identities on PARTI. Several members of the same family may share a Source account, each with his or her own PARTI identity.

Many of PARTI's regulars are known by several names. They use one name when they're all business, another name when they're involved in friendly or personal discussions. And there's no extra charge for having additional PARTI names.

Changing your PARTI account from single-user to multi-user is as simple as typing MODIFY ACCOUNT from the ACTION prompt.

Note, though—once you change from single-user to multi-user, you cannot return to single-user status again with that Source User ID.

In addition, when you change to a multi-user, each PARTI name connected to that user ID must have a password. So, as a multi-user account, you'll be greeted by several questions each time you visit PARTI:

Are you signing on to an existing subaccount?
What's your PARTI name?
What's your password?

Ordinarily, you would type YES to the first question. However, if you wanted to create an entirely new PARTI subaccount, you would type NO, then start the process of signing up the new PARTI name with the new password.

Changing to a multiuser account means more preliminaries are needed to enter PARTI, but you can string them all together from the -> prompt at command level, since you can anticipate the sequence of commands, as in:

****

-> PARTI,y,your PARTI name, your password <RETURN>

****

As a multiuser account, you don't have to log into PARTI anew for each "persona." Once you're inside the PARTI system, you can change from one PARTI identity to another with the BECOME command at the ACTION prompt.

Remember that all PARTI names listed under subaccounts are as separate as if they were logged in under different user IDs. Establishing a new subaccount is like coming into PARTI for the very first time. There's no crossover. Private messages sent to one PARTI name can't be retrieved by anyone with a different PARTI name on the same Source account. Each subaccount is secured with its own password which makes it perfect for multiusers of the same Source account.

A NOTE ABOUT COMMANDS

Throughout our PARTI discussion, we've typed out the complete command and put it in capital letters. Neither is necessary. Commands in PARTI usually can be abbreviated to the first letter with a few exceptions. When two commands begin with the same letter, such as FIND and FILE, they must contain enough letters to distinguish them from each other.

We suggest you spell the commands out each time until you become familiar with them. If your terminal is capable of generating small (lowercase) letters, you can use small letters and capital letters in commands interchangeably.

THE INVISIBLE PARTI

With all the activity among the public topics on PARTI, it's difficult to believe that the public topics and notes you see are just part (perhaps the smallest part) of the activity on the PARTI system.

There's much going on behind the scenes in private conferences. PARTI is a powerful and growing tool for businesses and individuals alike who need this form of electronic conferencing for more than simply pleasure and entertainment.

Private PARTI topics are used as well by universities for on-line seminars and continuing education.

In fact, the uses of the PARTI system are growing daily. We've known families who are separated by distances who keep in contact with each other through private conferences on PARTI. Groups of old friends who have left their hometowns for jobs elsewhere form private topics. As a result, when one of them writes a note, the whole group gets to read it. It's nothing less than a private, on-line newsletter that's delivered instantly to those eager to read it.

For you, PARTI holds the same possibilities—and a lot more. In fact, most people believe that the most important uses for this sort of conferencing have not yet been tapped. Electronic conferencing is still in its infancy. The structure is not the limiting factor. It's the hesitancy on the part of users to put it to more uses.

Reading this or any book, however, won't allow you to discover the full potential of the system. You have to see it to understand what it might mean for you personally.

A TOPICAL DISCUSSION

Logging into PARTI for the first time is akin to walking into a party that's been going on for a long time without you.

Let's face it. PARTI on The Source has been going on since 1982. You've missed a lot of ACTION. What you need is a little historical perspective—information about the public topics, past and present—that PARTIers have turned their attention to.

So what do PARTIers discuss? You name it! In this electronic conference hall, nearly anything goes. Posting a topic and connecting it to "PARTI" is sort of like getting up on a soapbox in the park and saying "I've got something I want to discuss here and I want everyone who's interested to join me."

Sometimes the depth of subject material is all but unfathomable. Often the computer-related discussions become so complex, you'd need an engineering degree to understand.

Other times, the topic notes sound like transcripts from group therapy. Recently, one PARTIer expressed his shock, outrage and complete puzzlement at finding out a trusted friend and business associate was arrested and found guilty of child molesting. He wrote six notes in a topic he started about the incident. The notes were straight from the heart. And there were others who joined his topic to offer what comfort they could and just let him know that someone was reading what he was saying.

We discuss some other interesting conferences in Chapter 16 when we share a few of our favorite things.

Following is a list of topics that existed on PARTI on The Source in late 1985. We can't guarantee that all of them still will be available. If you see one that piques your curiosity, you can find out if it's still around by typing READ and the topic name (in question marks) at the ACTION prompt.

Incidentally, we're not printing this list to advertise any particular conference or even to recommend them. We're printing it to show you the kinds of public topics that have been discussed, to give you an historical perspective, electronic style.

THE LIST

"RELATING—FOR YOU" One of the most popular of all personal topics on PARTI.

"ELECTRONIC LOVE" Sex, love and flirtation via the computer terminal.

"EATS" A whimsical conference about a culinary adventure.

"TELL MAMA" Mama says she is the "Dear Abby" of the electronic community.

"DREAMS" A topic to discuss, share and record dreams.

"PROFILES" PARTI personality autobiographies.

"HALLOWEEN" A Halloween PARTI as well as some serious discussion about Halloween.

"GAY RIGHTS" Discussions about all aspects of the gay life.

"INTERNATIONAL PARTI" A place for overseas PARTIers to make themselves known.

"NATL DATING BBS" A place to meet people.

"SMALL GODS" Called "A place for poets, dreamers and artists to share."

"DEATH AND DYING" Exploring emotions about death.

"E-MORALITY" A topic about morality and electronic love.

"QUIT SMOKING" A group effort to stop smoking.

"PLAYWRITING" A place for playwrights and theatergoers.

"EX-HIPPIES" What happened to the counterculture?

"SEX GADGETS" This one was featured on a Phil Donahue Show in the spring of 1985. Contains sexually explicit material.

"ROE V. WADE" About a Supreme Court abortion decision.

"DOE V. BOLTON" Another about a Supreme Court decision on abortion.

"FANTASY CONFERENCES" A discussion of fantasies. Conference contains sexually explicit material.

"TEEN ROOM" A topic for teens.

"WEIGHT WATCH" A dieters' support group.

"HOT LINE" A place to meet people "whether you're looking for a date, a chat or a good friend."

"CONSUMER PROTECTION" Uniting for strength in resolving grievances with manufacturers.

"SPLITTING UP" A discussion and support group about ending a relationship."

"CALORIE COUNT" Keep a log of what you eat.

"SUPPORT GROUPS" A structure for Support Groups.

"TOUGHLOVE" A new approach to teen discipline.

"STOP SMOKING" Support group to stop smoking.

"THE PARLOR" "A place for friendly conversation."

"PEOPLE AND PROBLEMS" Offer of services geared toward human communications.

"SINGLE PARENTING" A support group.

"CALLAHANS" Discussion of sex, marriage and love for men only. Contact KETZER for access.

"D&D CRIMES" The effects of role playing games on the youthful player.

"PARTI YEARBOOK" Biographies of those who contributed photos to the PARTI Yearbook.

"LATE BIOS" Biographies of those who didn't meet the deadline.

"AIDS NEWS" A conference about AIDS.

"SPACE" A must! This is a complete on-line "electure" dealing with the realities and philosophies of space travel.

"HOLOCAUST REMEMBERED" A discussion of the Nazi's Holocaust.

"BAN ABORTION NOW!" A topic for those interested in banning abortion.

"ALL MEN ARE JERKS" Not exactly an objective topic, some interesting opinions.

"WEDDING BYTES" Couples who met via The Source and who are either married or to be married.

"FORUM I" A place for new users to meet and receive tips about using PARTI.

"CP DISCUSSION" Consumer Protection Issues.

"WHO READS THIS STUFF" A look at feminism from a not-too-impartial observer.

"JAPANESE COLLEGE" A comparison of Japanese education to American. Also gives a glimpse into Japanese lifestyles.

"REAL NEWS" Organizer discusses "What's behind the stories that don't get major attention in the press?"

"TEEN ROSTER" A place for teens to exchange Source user IDs.

"ROSTER" A place for all the rest of us to exchange Source user IDs.

"SPACE ACTIVISTS" For "People who care about the future of space."

"APARTHEID" A discussion of the South African policy.

"PARENTING" A support group.

"THE BOMB" Thoughts on the nuclear age.

"SYSTEMS" Management information, executives on-line, office automation, decision support and telecommunications systems.

"TRANSFORMATION" Networking among business people who have "transformational" experiences.

"COPY PROTECTION" Technical and moral aspects as well as how-to discussions.

"TELECONSULTANTS" A topic for professional consultants.

"AGENTS" Subtitled "Everything you want to know about literary agents and then some."

"PHYSICIANS ON PARTI" An attempt at a PARTI medical network.

"LAN" Knowledge base for local area networks.

"COMPUTERS & LAWYERS" Computer uses in the law office.

"DBASEII" Topic for DBase II and other programmable database management packages.

"INTERIOR DESIGNERS" A topic for interior designers, related professionals and students.

"BARTER" Ways to leverage your business buying and selling.

"INSURANCE" A discussion of concerns, interests and experience with it. For consultants and small business people.

"PARTI IN PRIVATE" A public discussion of PARTI for private groups.

"CLIENT" Mike Greenly, a journalist and marketing consultant, tells what it's like to launch a new piece of software for a client.

"TELECOMMUTING" A discussion of commuting via computer and telecommunications.

"EMAIL V CC" Comparing electronic mail and computer conferencing.

"SALES PARTI" Sales and sales management applications of PARTI.

"NETWORKSHOP TRAINER" Teleconsultants who offer on-line training.

"PLANNING PRACTICES" Current practices and trends to link corporate and information systems planning.

"GODZILLA MEETS BAMBI" How big computer corporations treat noncorporate users.

"ACCOUNTING SW INFO" Accounting software for micros.

"CBT" Computer-based training.

"SMASH IBM?" The topic is "Is IBM getting dangerous?"

"SYMPOSIUM II" Social and organizational implications of computer conferencing.

"APPLE II" PARTI users group for Apple II and Apple //+.

"APPLE //c" A discussion of the computer at the time it was announced.

"APPLE OBSERVATIONS" A place to participate in observations about Apple computers.

"MACINTOSH NOW" A place to discuss the Macintosh.

"APPLE STOCK" Market analysis of Apple stock.

"DOCUMENTATION" A discussion of user manuals.

"LOTUS" A topic devoted to Lotus for the Apple.

"NEW SHELTER SOFTWARE" Free software from *New Shelter Magazine*.

"WORDSWORTH" Finding the right word processing tools.

"MODULA-2" A users group for Modula-2.

"MAC SOFTWARE CENTER" Public sharing of information and experiences with software for Apple's Macintosh.

"POPULAR COMPUTING" An electronic edition, updated weekly, of *Popular Computing* magazine.

"ASK POPULAR" Here's where you can ask questions of *Popular Computing* magazine.

"GOODIES" Short, newsy and juicy gossip about the computer industry.

"ASCII EXPRESS" A discussion of this terminal program for Apple computers.

"JAZZ BRIEFING" Lotus' Jazz for the Macintosh.

"IBM OBSERVATIONS" A place to discuss IBM computers.

"ASHTON-TATE" A discussion of DBase II and other database software.

"TURBO" A discussion of Borland's Turbo Pascal.

"MODEMS" What to look for, whether to buy 1200 or 2400 baud.

"HOMEBASE" A discussion of this IBM PC program.

"ABILITY UG DIGEST" Information about the Ability User's Group.

"ENVIRONMENT" Talk of environment managers for the IBM PC—TopView, MS Windows, and GEM.

"IBM PC" PARTI users' group for the IBM PC.

"WORDSTAR/ASCII" Program to convert Wordstar(tm) files to and from ASCII.

"TELENET" The topic is the quality of Telenet's interconnect to The Source.

"VICTOR 9000" A place to share information about the Victor machine and its software.

"ADS" A discussion of local bulletin board systems.

"PORTABLE COMPUTERS" Covers lap and portable computers.

"COMPUTER KIDS" Seeking your opinion about youthful pirates and hackers.

"FAMILY & COMPUTERS" The impact of computers and related technology on family life.

"OFF LINE" Reducing costs and improving useability of The Source through off-line processing.

"HANDICAPPED" Computer applications for the handicapped.

"FREEWARE" A new concept in software distribution.

"BLINK ME" PARTI regular Steve Gibson has come up with a way to cut on-line time radically by writing notes off-line and "blinking" them quickly into the PARTI system, once you're on-line. He says this system cut his on-line time by 90 percent. It's worth reading and using.

"SINCLAIR INFO NET" For Timex and Sinclair users.

"WHAT WE DO FOR MONEY" Thoughts of a computer trainer.

"SERVICE RIPOFF" The high cost of computer repairs.

"TAX LOGS" An automated process to document your computer use.

"MODEM TIMES" Updated numbers of BBSes.

"C" The C Programming Language.

"FUTURE WORLD" Impressions from the Future World Expo in San Francisco.

"COMPUTER SOUL" The impact of computers on ethics.

"UNERASE.ASM" A program to get back erased files if you erased them under the CP/M operating system.

"RIP VAN WINKLE" A professional programmer returns to the industry after a long hiatus and criticizes modern personal computers.

"SMALL COMPUTERS" How efficient is your personal computer's software? As the trend goes to larger machines, do we see the quality of software deteriorating?

"ENA BULLETINS" Bulletins from the Electronic Networking Association.

"ENA FORUM" General conversation about the Electronic Networking Association.

"ENA MEMBERSHIP" Membership form for the Electronic Networking Association.

"ENA" Anchor conference for the Electronic Networking Association activities on The Source.

"ABORTION" A study of morality and permissibility of abortion under various circumstances.

"EASTERN CATHOLICS" A history of origins and rites of the Catholic Church.

"SKEPTICS" A discussion of theories labeled as occult and paranormal.

"CHRISTOLOGY" An on-line course in Christology.

"CHURCH" A discussion of various elements of the Catholic Faith.

"LITURGY" A study of the Rubrics and meaning of the Liturgy of St. John Chsysostom.

"PRAYER" A study of the nature and practice of prayer as found in Eastern Christianity.

"INNER VISION" A discussion of mysticism.

"HACKING—FOR STEW" The ethics of "hacking."

"SCRIPTURES" A discussion of the New Testament.

"WOMEN IN CHRIST" Scriptural roles of women.

"RELIGIOUS ASSOCIATES" How religions may use PARTI to support their networking efforts.

"CHURCH SURVEY" Computer networking and churches.

"PILGRIMAGE" A forum to share religious beliefs.

"SPACE AND SPIRIT" What are the spiritual and religious issues raised by space exploration?

"TIME" Can we move in time other than at our usual pace?

"POLITICS" A forum for political discussion.

"LEGISLATIVE EMULATOR" A discussion of current topics facing the U.S. legislature.

"AMERICA" American heritage: Its past, present and future.

"GOP 84" An experiment from the Transcontinental Electronic News Service, live from the 1984 GOP National Convention.

"KENNEDY" The life and death of JFK.

"LOCAL POLITICS" The premise is "You can make a difference on the local level."

"STAR WARS" The "Star Wars" defense system.

"AMENDMENT II" The right to defend yourself.

"WAR AND PEACE" How to attain peace in the nuclear age.

"AMENDMENT IV" Concerning illegal search, seizure and arrest.

"SPACE DEFENSE" Space-based laser defenses against nuclear missiles.

"SPACE TAX" A tax for funding space programs?

"TENANT" Organizing a tenant's committee.

"DIVESTMENT" Stock holdings in South Africa.

"TRIBLE BILL" Will an attempt to suppress child pornography result in restrictions on computer networks?

"POST SURRENDER" What happened immediately after Japan's surrender at the end of World War II?

"HOUSE CALLS" Answers to medical questions for use in fiction. The topic is intended to lend credibility to writing when it comes in contact with the medical field.

"DOMESTIC ROBOT" A discussion on the possibility of robots for housework.

"BEASTIES" Do sea monsters, dragons and dinosaurs really exist?

"ALTERNATIVE VISIONS" A discussion of positive alternatives to nuclear war.

"2020" A look into the future.

"UNIVERSE" A discussion of time, space, universe and gravity.

"SPACE RORSCHACH" "Your innermost visions of outer space."

"SPACE BEINGS" Who and where are the extraterrestrials?"

"RESEARCH SYSTEMS" How we acquire and use knowledge, information and data for science and technology.

"PIE-IN-THE-SKY" "Chasing universal untruths."

CHAPTER 12

Getting Personal with The Source

There are a lot of ways to communicate, from letters and phone calls to a rendezvous in the park. The most intimate way is to invite someone into your home. In a sense, you can do that on-line, too

We've been telling you that the User File Directory is your private area of the system. Also, we've hinted that there's a public side of the UFD as well. In an area called "Sharefiles," you can place articles, newsletters, even public-domain programs and invite your electronic neighbors to enjoy them. And, of course, it works both ways—you can visit Sharefiles offered by other subscribers.

In this chapter we'll hop on-line for a quick tour of Sharefiles. You'll see how to set up your own public area and then look at a brief file left on Charlie's "front porch."

After that, we'll log off and regroup back here to talk about how you can use this special part of the system to publish your own newsletter.

When we first talked about The Source back in Chapter 1, we observed that the system grew, in part, out of a response to the idea of electronic publishing. The romantic notion was that anyone with a modem could have an on-line magazine. Many of the earliest settlers on this system were interested primarily in sharing words and programs electronically.

The way The Source has evolved to meet those needs gave rise to the Member Publishing section you saw in Chapter 7. Perhaps you've explored this neighborhood system on your own by now. All of the Member Publications began—and still begin—as humble experiments in the Sharefiles areas. Using the procedures described in this chapter, you'll be able to easily set up a public, menu-driven section of your UFD that will look just like the Member Publishing newsletters you've seen. And who knows—if you come up with a particularly popular feature, The Source may want to pay you a royalty to continue publishing it.

Then, in the second half of this chapter we'll tell you how to get intimate with the system in another way—by transmitting and receiving files directly between your microcomputer and The Source.

This procedure, called uploading and downloading, goes hand-in-hand with creating public files. No one wants to pay for the connect time to stay on-line long enough to write a thought-provoking two-thousand word article. Obviously, it would be smarter to write it off-line with your word processor and then upload it to The Source. We'll show you several ways to do that.

And those same procedures let you write other things off-line—letters to be delivered to the SourceMail program, notices for POST, or messages for PARTI—virtually anything that you write for the system.

In other words, hang around, pal. . . . We're talking about saving The Big Bucks.

CONNECTING TO THE SOURCE TO SHARE THE WEALTH

To see how Sharefiles work, we first need to be a visitor. Then we'll learn how to be a host. So we'll start by reading the file that Charlie left for us.

Log on to the system following the usual procedure and turning CHAT -OFF. Then at the Command Level prompt, type:

****

–> TYPSHR <RETURN>

****

This command tells the system, "I want to TYPe a file from someone's SHaRefiles area." The Source will ask you to identify the area you want to visit, that is, to enter the user ID of the owner.

****

From what user's sharefiles area?

****

Charlie's user ID is BBZ824, so type in that number at the prompt, and the system's next request will be:

****

Enter file or pathname

****

In other words, what's the name of the file in BBZ824's Sharefiles in which you're interested?

Type THISFILE <RETURN> and the system summarizes the command it's executing:

****

TY SFILES>BBZ824>THISFILE

****

Then it displays the file you requested. It'll just take a few seconds to read—certainly not some of Charlie's more deathless prose.

As that file points out, you can copy public files as well as read (or "TYpe") them. Right now, if you were to type LD <RETURN> to List the Directory of your UFD, you'd not find a copy of THISFILE. That's because TYPSHR only displays the file, rather than copying it. If you wanted a copy of it, you could enter a different command—GETSHR (for "GET a file from a SHaRefiles area").

Let's do that. At the -> prompt, type GETSHR <RETURN> and deal with the prompts as we indicate here:

****

-> GETSHR

Get file from what user sharefiles area BBZ824 <RETURN>
Enter file name THISFILE <RETURN>
Enter new name or c/r <RETURN>

****

In a moment, the system reports:

****

COPY SFILES>BBZ824>THISFILE

****

It then returns you to the -> prompt.

Now if you enter LD, you'll find that THISFILE has been placed in your own UFD, just as if you had typed it there yourself. And, as with any file in your UFD, you can edit it, alter it, mail it to someone else, etc., or just read it again by entering TY THISFILE <RETURN>.

If you want to try any of those things with THISFILE, do it now. When you're finished, delete the file by typing DEL THISFILE <RETURN> at the -> prompt. (There's no use paying storage costs for it.)

The lesson from this little exercise is that you can use the TYPSHR and the GETSHR commands to retrieve files from any other user of The Source, as long as you know (1) the user ID of the subscriber who's doing the sharing, and (2) the name of the file being shared.

Also, take note of the style of that summary report The Source has displayed twice now:

****

TY SFILES>BBZ824>THISFILE

****

And:

****

COPY SFILES>BBZ824>THISFILE

****

That's a kind of shorthand that will be useful to you as you become more familiar with Sharefiles. You can type it as a special command string at the –> prompt to sidestep all those subsequent prompts that came with TYPSHR and GETSHR.

OPENING YOUR OWN FRONT PORCH

Suppose you'd like to open a section of your file area to host electronic visitors. To demonstrate that, we need a test file, so at the –> prompt, type:

****

ENTER TEST <RETURN>

****

Now type in several lines of text. If you're at a loss for words, we'll suggest some. (The wording isn't important since we're not *really* going to make this public. This is just for practice.)

****

This is a test of the new Sharefiles area. <RETURN>
I'll be moving this file to my shared space, <RETURN>
where others could retrieve it, if they <RETURN>
knew (1) under which user ID it was placed <RETURN>
and (2) the name of the file. <RETURN>
cheers, my name <RETURN>
<RETURN>

****

Very good. Of course, if now you were to enter an LD, you'd see that TEST is in your UFD.

To move TEST to the public area, you need to open your Sharefiles section for business. (Remember back when you were working with SourceMail and you had to open a filing cabinet to store mail? This is the same sort of thing; you need to tell the system you want to create a Sharefiles department.)

At the –> prompt type

****

-> SHAREFILES <RETURN>

****

and the system gives you a message like this one:

****

WELCOME TO THE SOURCE SHAREFILES SYSTEM BBP112

Your shared files area has been created.
To place files into it, use the COPY command
at command level. To get into this area, always
use the sequence :

-> A SFILES>BBP112

Or:

-> ASF

This is where your shared files will be stored.

****

(Of course, the message will include *your* user ID, rather than Dave's.)

Notice that the message speaks of "getting into" this new area. You'll find it's helpful to think of Sharefiles as a kind of separate "room" added to your UFD. To use it, you need to know how to move from your main UFD to the Sharefiles section.

To get a clearer picture of where you "are" now, type WHO <RETURN> at the -> prompt and The Source will display something like this:

****

<USR18B>BBP>BBP112

****

That's the WHO message when you're in your regular UFD. Let's see how it looks when you're "attached" to Sharefiles. Type:

```
******************************************************************

    -> ASF <RETURN>

******************************************************************
```

When you receive the -> prompt, type WHO <RETURN> again, and you'll see something like this:

```
******************************************************************

    <(DBAS06>SFILES>BBP112

******************************************************************
```

The phrase "SFILES>" in front of your user ID means you've moved from your regular UFD area into Sharefiles.

To move back to your UFD, type:

```
******************************************************************

    -> ALC <RETURN>

******************************************************************
```

That attaches you to your regular (private) file area again. (And, to make sure, you can type WHO <RETURN> and you should see the correct file report.)

So ASF and ALC are a pair of related commands. The first takes you from the UFD to SFILES. The latter takes you from SFILES back to the UFD.

To copy a file into your Sharefiles area, we'll need a new command. It's PUTSHR. The file must already exist in your regular UFD. You might think of your UFD as a work area; Sharefiles is for the finished product.

Let's put TEST into the Sharefiles section. At the prompt, type:

```
***************************************************************

    -> PUTSHR <RETURN>

***************************************************************
```

and answer the prompts:

```
***************************************************************

    Please enter filename TEST <RETURN>
    Please enter new filename or c/r

***************************************************************
```

Hold it a sec. You saw this prompt earlier when you copied Charlie's THISFILE into your UFD. That time we just pressed the RETURN key at the prompt—that told The Source to copy it with the same file name that Charlie used. However, as the prompt indicates, you may rename the file as you're copying it, if you'd like.

Let's do that. At the latest prompt, type MYTEST <RETURN>. Now the instructions are summarized by The Source as:

```
***************************************************************

    COPY TEST SFILES>BBP112>MYTEST

***************************************************************
```

Translation: "Copy the file called TEST from my UFD to my SFILES area, renaming it MYTEST."

You could take our word for it that the file has arrived safely, but perhaps you should look for yourself. Attach to your Sharefiles area (type ASF) and then enter LD to List the Directory there. You should see one file—MYTEST.

KEEPING IT CLEAN

The operative word in all this sharing business is "copy." By placing something in Sharefiles, you haven't *moved* it at all; you've made a copy of it. At the moment, then, you have two versions of your work: TEST in your UFD and MYTEST in your SFILES area.

To save money on storage costs, you can routinely delete the original from your UFD once you've placed it in SFILES.

By the way, all the commands you've learned for your UFD also will work in your Sharefiles section. So SFILES text still can be changed with any of the editing commands you've learned. Just attach to your SFILES and type ED followed by the file name, as in ED MYTEST. When you're finished, enter FILE <RETURN> as always to save the edited version.

In this case, since TEST and MYTEST both are just demonstration files, there's no good reason to keep either one of them. Therefore:

1. While you're still in your SFILES area, type DEL MYTEST <RETURN>.
2. Then attach to your regular UFD (type ALC).
3. After you've arrived (that is, after you see the -> prompt again), type DEL TEST <RETURN>.

Now log off as usual. We want to put some of this new information in perspective.

SPREADING THE WORD

Nothing too hard about all that, was there? It's just a matter of remembering that every Source subscriber can have *two* file storage areas: one private (the UFD) and one public (Sharefiles). The UFD can serve as a work area from which files can be submitted to SFILES. Then anyone on-line who knows your user ID and the name of the shared files can read them or copy them. And you can visit other users' SFILE areas.

But, how do you find them?

Scanning the POST bulletin board is the best way. The Source has established a special category for users wanting to spread the word about their public offerings. It's called PUBLIC-FILES. We usually scan POST with a keyword of PUBLIC, as in:

****

-> POST S K PUBLIC <RETURN>

****

Of course, you also could Scan or Read individual categories, instead of using a keyword.

So, if you have a file you want to share with fellow Source travelers, be sure to leave a message on POST to tell them about it and mention the user ID and the file name they'll need to know to get access to it. In fact, your notice can even include the specific command string we've just learned, telling them something like, "Just enter TY SFILES>BBZ824> THISFILE at the Command Level prompt."

Some people use Sharefiles as a testing ground for Member Publishing ideas. Before taking the plunge into a full-fledged newsletter, they give the idea a test audience for a month or two in SFILES. That way they can gauge whether there's sufficient interest to make it a workable proposition. Usually, these prospective publishers tell you in their POST messages that they'd like feedback from you. On-line etiquette suggests that if you like what you see or have ideas for improving it, you should drop the author a SourceMail about it.

Some of the sharing members, especially those with electronic publishing ambitions, will offer a pretty sophisticated package in their SFILES area, complete with menus and letters-to-the-editor sections, just like those you see in the Member Publishing area. Their POST messages usually contain a different command string for access, such as:

****

PUBFILE BBZ824

****

And that brings us to how you, too, can get in on the ground floor of the electronic publishing game.

BECOMING AN ELECTRO-PUBLISHING MOGUL

When we toured Member Publishing in Chapter 7, we made an observation that, while The Source gives its members lots of leeway in how they present their newsletters, the following style of menu is quite popular:

****

FILE NAME	LIST TIME	TIMES READ	LAST READING	
1 HELLO	0:43	471	85–03–04	10:46:11
2 THIS WEEK'S THOUGHTS	1:23	941	85–03–04	08:33:09
3 LAST WEEK'S THOUGHTS	5:32	1599	85–03–05	10:25:02
4 NEXT WEEK'S THOUGHTS	3:34	452	85–03–05	02:04:04
5 THE MAIL BAG	3:36	182	85–03–03	23:20:33

You may enter a list of options separated by spaces.

Enter list of options, RETURN for menu, <H>elp or <Q>uit:

****

There's a good reason for its popularity. This style of menu, and the free on-line software that makes it work, are provided by The Source. It's called PUBWARE and you don't have to be a Source official or even an entrepreneurial Member Publisher in order to use it. Right now, you can use PUBWARE to set up a menu-driven presentation of your own Sharefiles. The software will even allow you to keep a running log of the user IDs of your readers, a great way to find out if your files meet a need in the electronic community.

The menu software was designed for those who would like to be included in the Member Publishers row (an area that we also called PUBLIC, you'll recall). An active readers' log is the way The Source determines if a prospective publication makes the grade.

However, even if you don't have ambitions of a publishing empire, you still might want to use the free PUBWARE package to make neat displays for the friends and associates who frequent your Sharefiles. For instance, some business associates who use The Source for on-line communications might want to set up a menu-driven file area to make it easier for all to find the shared information.

Either way, all the public programs you need are available from a special Source account called "STCPUB" and they use a pretty clever command (CO) file, just like the ones you've written. All in all, it will take only about fifteen minutes' worth of connect time to get the software and set up the menu display.

YOUR LIFE IN ROCK 'N' ROLL

To see how to get and use PUBWARE, let's pretend you're going to turn your Sharefiles into a publication about music, fulfilling your lifelong ambition to be a rock music critic. Your plan is to write three or four articles a month—reviews, commentaries, biographies, anything on your mind—and post them on the system for the world to share and comment on.

While you're doing this primarily for the love of the game, you're not ruling out the possibility of making some money at it. In other words, if your commentaries really make a splash, you're thinking of approaching The Source with the proposal that your newsletter be moved "uptown" to the PUBLIC area. (You could use the royalty payments to offset your connect-time bills, anyway.) However, all that's in the future. First, you'd just like to develop an audience for old-time rock 'n' roll.

First, then, you write the articles for your premier issue and store them in your Sharefiles area, just as we did with TEST during the tour. (In the last part of this chapter, when we talk about uploading, we'll have some time-saving tips on that.)

After the articles are on-line, you need to get the PUBWARE programs to put them on a menu. You start by going to your Sharefiles area—type ASF <RETURN> at the prompt. Then you use the GETSHR command like this:

```
************************************************************

    -> GETSHR <RETURN>

    Get file from what user sharefiles area STCPUB <RETURN>
    Enter file name PUBWARE <RETURN>
    Enter new name or c/r <RETURN>

************************************************************
```

As you learned on the tour, this command sequence copies a file called "PUBWARE" from STCPUB into your Sharefiles.

PUBWARE actually is a CO file written by officials at The Source and saves you time in copying several other files from STCPUB. All you have to do is type:

****

-> CO PUBWARE <RETURN>

****

Now the system automatically executes the GETSHR command five more times, fetching files called MENU.MAINT, MENU.FILE, READ.LOG, HEADER, and END.PUBWARE.

At the end, you see this message:

"The transfer from STCPUB is now complete. At -> below, type DEL PUBWARE and press RETURN. Then type LD to show the list of files in your Sharefiles area."

That's saying that since the CO file has done its job and copied the files, you now can jettison it. (You can delete the END.PUBWARE file as well, since it's just the text file that contained the closing message you saw.)

MAKING THE MENU

Four files remain—HEADER, MENU.FILE, MENU.MAINT, and READ.LOG—and they're the ones that will build and operate your menu. MENU.MAINT and MENU.FILE are BASIC programs, and we need to "run" the first one to get this show on the road.

So type this command:

****

-> BASICV MENU.MAINT <RETURN>

****

This tells The Source to summon its BASIC language (called "BASICV") and run the MENU.MAINT program. After a moment, you'll see:

****

```
MENU.FILE MAINTENANCE PROGRAM V3.4
(Enter ? for info about commands) . . . Please wait
# READ NAME LTIME LAST READING
(A)DD, (C)HANGE, (D)ELETE, (T)IME, (S)AVE, (Q)UIT?
```

****

This program is going to build a menu around your music articles, all of which also are in the SFILES area. You need to begin by typing A <RETURN> to add items to the menu. The program prompts you for the names of each file you want to appear on the menu:

****

```
New file names to add(hit return for menu)
#1?
```

****

And you might enter BOB.DYLAN if that were the name of the first review. You would continue entering the names of articles until all were on the menu. Then just press RETURN on a blank line when you're finished.

The program then constructs the menu and displays it, such as:

****

```
#  READ  NAME              LTIME  LAST READING
1    0   BOB.DYLAN          0:00  85–05–04 11:58:10
2    0   CHUCKS.CHILDREN    0:00  85–05–04 11:58:19
3    0   PRINCE             0:00  85–05–04 11:58:23
4    0   DA.DEAD            0:00  85–05–04 11:58:27
5    0   REACHING.US        0:00  85–05–04 11:58:31

(A)DD, (C)HANGE, (D)ELETE, (T)IME, (S)AVE, (Q)UIT?
```

****

Notice that the file names appear on the menu, so you'd want to be careful to give them descriptive names. The CNAME command (for Change Name), which we learned earlier, would be handy here if you wanted to alter a file name before putting it on the menu.

When you've got the menu the way you want it, you can "time" each article. (You saw in the Member Publishing newsletters that the menu gave an indication of how long it would take to read each article at 300 baud. That service is a function of the MENU.MAINT program.)

If you select T, the system gives you an opportunity to time all of the articles or selected ones, and then displays the findings:

```
**********************************************************************

BOB.DYLAN                                   List time= 4:21
CHUCKS.CHILDREN                             List time= 3:04
PRINCE                                      List time= 2:22
DA.DEAD                                     List time= 1:55
REACHING.US                                 List time= 0:34

(A)DD, (C)HANGE, (D)ELETE, (T)IME, (S)AVE, (Q)UIT?

**********************************************************************
```

The S option saves the changes; Q quits without saving the changes.

Once the menu is created and saved, you can change it by just running the MENU.MAINT program again and selecting the <C>hange or <D>elete options.

DESIGNING YOUR COVER

The HEADER file contains your title page. It's like any other text file you'd meet on the system. Just enter:

```
**********************************************************************

-> ED HEADER <RETURN>

**********************************************************************
```

Then tap RETURN to go immediately into INPUT mode. Type away, perhaps something like this:

****

Eddie's Eclectic <RETURN>
Rock Garden <RETURN>
-rock 'n' roll, now 'n' then- <RETURN>
Enjoy! <RETURN>
(and tell me what you think!) <RETURN>

****

At the bottom, you press RETURN on a blank line to get back into EDIT mode, then type FILE <RETURN> to save it.

Now you can see how the whole thing looks by first going back to your regular UFD—type ALC <RETURN>. Then, at the -> prompt, type in PUBFILE followed by your own user ID number, as in:

****

-> PUBFILE BBZ824 <RETURN>

****

The system will display your publication, just as your readers will see it:

****

MEMBER PUBLISHING

Version 6.0
(C)opyright Source Telecomputing 1985
Logging BBZ824. . . .

Eddie's Eclectic
Rock Garden
-rock 'n' roll, now 'n' then-
Enjoy!
(and tell me what you think!)

FILE NAME	LIST TIME	TIMES READ	LAST READING
1 BOB.DYLAN	4:21	0	85–05–04 11:58:10
2CHUCKS.CHILDREN	3:04	0	85–05–04 11:58:19
3 PRINCE	2:22	0	85–05–04 11:58:23
4 DA.DEAD	1:55	0	85–05–04 11:58:27
5 REACHING.US	0:34	0	85–05–04 11:58:31

You may enter a list of options separated by spaces.
Enter list of options, RETURN for menu, <H>elp or <Q>uit:

****

If you'd like, you could look over each article on the menu and read the Help files (which come as a bonus of the PUBWARE programs). When you're done, a Q at the prompt returns you to the –> prompt.

After that, you should clean up the work area. You'd probably want to delete the originals from your UFD once they're secure in your SFILES section.

Incidentally, the READ.LOG file, which keeps tabs on the user IDs of your readers, is optional. You can run the program without it, if you don't want a log. However, once it's active—that is, once you've used the MENU.MAINT program to create the menu—never delete or alter the READ.LOG. That could cause your publication to be inaccessible. So, if you decide you don't need a reader's log function, it's a good idea to delete READ.LOG before you create the menu.

READ.LOG, which is just an ordinary text file that can be viewed with the TY command, can grow rapidly once you start telling your on-line acquaintances about your publication. In a minute, we'll have a tip for managing that file.

HAWKING YOUR WARES

Now you've got everything on-line for your debut as a rock music commentator. There are two things you need to do to "go public."

First, you need to notify the people at The Source that you're using a PUBWARE display and let them open the access level to your Sharefiles area. (This isn't necessary for the simple sharing we saw with the TYPSHR and GETSHR command. It's only if you're using the menus created by PUBWARE that you need to contact The Source people.) So, write a short SourceMail to STCPUB with the subject of "Access." You can take the fast lane with this command string:

****

-> MAIL STCPUB SU ACCESS

****

They will acknowledge by return electronic mail and pass the message on to the technical staff. The door of your SFILES publication will be opened in a day or so.

Second, when you're open for business, you need to let the other Source members know. (So sad to give a party if nobody comes.) As you've seen, the best way to advertise your new baby is through POST. You might place a couple of notices on different categories. Be sure to have at least one in the PUBLIC-FILES category. Each message should include, of course, the direct address of your publication—PUBFILE and your user ID. And don't forget the resources of the Member Directory for finding people interested in your project. You could search for those interested in MUSIC, ROCK, POP, etc. You might even want to build a distribution or mailing list for reaching them. See our discussion on that in the On-line Survival Kit.

COST CONSIDERATION: CAN YOU AFFORD TO MAKE THE BIG TIME?

Publishing the Rock Garden in Sharefiles isn't going to make you rich. When you publish it on your own, you *don't* get a percentage of your readers' connect time. You get royalties *only* if your publication later is accepted for the PUBLIC area of the system.

In fact, it will cost you money to open your Sharefiles. You can check the RATES feature on-line for the latest cost of file storage. Generally speaking, it would cost about $9 a month in storage fees if your publication had ten menu options, each with an article of about 4K (plus the necessary program files we've mentioned). Add to that the cost of the connect time needed to put the articles on-line and to write the POST messages and follow-up SourceMails for promotion. As a rule, most SFILES publishers spend about $40 a month for all expenses, including connect time, according to people at The Source.

Can you turn it into a profit-making venture? Well, you probably won't be buying a sailboat right away, but if your SFILES publication is received well enough by your readers, you might be able to convince The Source to take a chance on you. After a month or two of test marketing, you could send SourceMail to the STCPUB folks and get the talks started.

They'll want to know about how your readers' log has shaped up and what you've done to promote your publication on your own, through POST and SourceMail. If they accept your product for the PUBLIC area, they'll pay you a percentage of the connect time generated by your publication. That would mean there's an excellent chance that it will at least pay for itself. And more than likely, you'll have some bucks left over as profit.

CUTTING COSTS

While waiting for the ship to come in, there are things the aspiring electro-publisher can do to save some on-line money.

For instance, write your on-line articles tightly. Not only will this save you storage costs, but it will make your readers happy; they won't feel as if you're wasting their connect time. And happy customers are repeat customers.

Keep your work area cleared of unnecessary files. Once you've got everything in place in the SFILES section, there's probably no need to have duplicates in the regular UFD, so delete them.

But don't get overzealous in your housecleaning. While you *can* get by without the READ.LOG file, you'll need it to document your success if you want to eventually get on the Member Publisher's row in PUBLIC.

The READ.LOG file can grow rapidly if your publication catches on, so it's good to know some ways to manage it.

Remember, *don't* delete or change READ.LOG's name once you've set up your menu. Instead, it's wise to open it from time to time and transfer the contents to another file. To do that, we use a new command—DUNLOAD.

The Source suggests that you might want to save your READ.LOG file records on a weekly basis, so you could type:

```
******************************************************************

-> ASF <RETURN> (attaching to Sharefiles)
-> ED READ.LOG <RETURN>
EDIT
N2 <RETURN> (to move the point down two lines)
DUNLOAD WEEK2 <RETURN>
FILE

******************************************************************
```

The DUNLOAD command is used to move the contents of one file to a new one. In this case, it has cleared out READ.LOG, saving the contents, to a new file called WEEK2. Now READ.LOG can begin collecting new material.

Finally, the very best way to cut on-line expenses is to be an earnest uploader/downloader. And that brings us to the second major topic of the evening.

THE UPS AND DOWNS OF FILE TRANSFERS

We've just demonstrated how you can get a message into your Sharefiles area by typing it directly into the system. But that's not how the "old pros" do it. What you read from others on The Source often originates off-line and then is transmitted to the system at a great savings in time for the writer.

In fact, we wager that once you're familiar with the techniques for *downloading* (getting stuff *from* the system) and *uploading* (sending stuff *to* the system), you'll never see The Source in the same light again. Instead, you'll be more inclined to use your word processor to write your Sharefiles texts or SourceMail or POST and PARTI messages, save them on a disk, then connect to The Source and deliver them at lightning speed; and to zip through the system like a sprinter, picking up long files as you go—everything from user instructions to interesting articles and news accounts—to save on a disk or a printout for reading after you've logged off.

The Source has a number of utilities to let you handle these tricks, if your computer and communications program can take care of business on your end. No guarantees. There are just too many different kinds of computers and programs out there to safely claim that all of them will cooperate with The Source on such an intimate level.

Sound familiar? That's the same plea we copped back in the first chapter as a reason for not talking about specific computers and programs.

However, now that we've hedged all our bets, we can say some *general* things about what you need in order to use these electronic dump trucks to haul to and from The Source.

Let's take a breather while you fetch the user's manual of your communications program.

Meanwhile, if you're already a well-heeled traveler to whom uploading and downloading are just old hat, take a walk. You can catch up with us in a couple of pages at the heading, "On the Upbeat."

GETTING YOUR GEAR TOGETHER

Let's start off by allaying a few apprehensions.

First, if your computer has disk drives and you're using a modern communications program, you probably can use at least some of The Source's upload and download features.

Second, it's not hard. This, despite the fact that some people write user manuals in the dialect they used back home on Neptune.

Most good communications programs have one or two ways to transmit files to and receive them from a remote computer system like The Source.

The first way to upload and download is used most often with text files—like letters, articles, messages, etc. The notion is something this:

1. You write a letter with your word processor and save it on a disk.
2. You run your communications program, log on to The Source, and go to the area where you want to send the file (such as the SourceMail program, PARTICIPATE, your User File Directory, etc.). You get everything ready, just as if you were going to type the message in manually (opening the file, etc.).
3. You tell the communications program to load your letter from the disk into the memory of your micro (it goes into a "buffer"). Then you enter a command to tell the program to transmit the letter from the buffer to The Source. (From The Source's point of view, it's just as if you *were* typing the file in manually, though The Source thinks that suddenly you sure can type fast.)
4. When the transmitting's all done, you close the file on the system as always.

It's step 3 where all the variations come in. Your communications program may have several ways to transmit: the whole document at once in a bulk transmission, and/or one line at a time (with a brief pause between each line or perhaps it waits for you to tap the RETURN key for each line), perhaps in a bulk but "formatted" transmission (with the program sending a carriage return on each transmitted line before the eightieth character or so).

For downloading, it's the same idea, except that you use a command to tell your micro's program to open the buffer and "capture" whatever appears on your screen from that point on. Your program might automatically save the incoming material on disk, automatically print it out, or just hold the material in the buffer until you specify what you want done with it.

Most of the confusion in talking about all this high-tech stuff comes, in part, from the lack of standardization in the terms. In your user's manual, for instance, this kind of transmission might be called, oh, "character dump," "buffer dump," or "ASCII dump." We've even seen the ever-popular "non-error-checked file transfer."

Well, don't let the Neptunese drag you down. There are just two basic characteristics of this kind of sending and receiving:

First, the files you upload and download are in "ASCII" format, the native tongue of The Source. (Most word processors already save files in ASCII, or at least can. Check the manual that came with your word processor to see if it routinely saves in ASCII. It may need a special instruction to do it. The bottom line is that the files you send to The Source must be in ASCII.)

Second, the files are sent and received without being checked for errors.

Beg pardon? What's this error-checking stuff?

That leads us to the second major kind of uploading and downloading, usually called "binary transfer" or "error-free transfer." It's often used for sending and receiving programs. But to get a handle on that, we need to backtrack.

A moment ago we said that the first method often is used for transmitting and receiving *text* files—letters and articles. In a word, *words*. Neither your computer nor The Source makes any effort to check and verify that every character you sent was correctly received.

That makes sense. Now that you've been around the block a few times on The Source, you know that from time to time, funniness happens on the telephone lines. A letter or two gets messed up because of a line glitch. On most text files, it's not a big deal if *the* comes out as *tje* or *United States* is received as *Uxited States*.

On the other hand, if you've ever done any programming, you know that one misplaced byte can make the program crash. That's why the ancient wizards of these telecommunications communities have come up with several different ways of getting around telephone-line noise.

All the methods use slightly different procedures, but they all work about like this: the sending computer transmits some material; then it and the receiving computer compare notes to see if what was received is the same as what was sent. If not, that block of data is sent again. If the material is the same, then the sending machine moves on to the next block of information.

You may have one of those error-free file transfer programs in your communications package. You may be able to use it with The Source. However, we'll wager that of the two, the first method will be the one you use most often.

STOPPING AND STARTING—"FLOW CONTROL"

That takes us to one more point—something the wizards call "flow control."

While you're sending material from your micro, The Source is gathering it as quickly as it can and putting it into a buffer, similar to the buffer in your own software. All goes well, unless your transmission gets a little ahead of the receiver. When The Source senses you're about to overflow its buffer, it needs to send a "hold-up-there-buddy" signal to your computer.

Think about the CONTROL S we've been using to freeze the display on our tours. This is the same idea. The Source sends a CONTROL S to your machine to say, "Wait—gotta empty some of the buffer." When it's ready to receive more, it sends your micro a "Come on" signal (CONTROL Q) and they continue talking.

Just about all modern communications programs recognize this flow-control procedure for sending and receiving ASCII files. Your manual should say if yours does (though it might call it by some intergalactic name like "XON/XOFF protocol," "DC1/DC3 standard," even "freeze/unfreeze recognition."

And that, friends, is as technical as we need to get. And it should put us all on common ground from which to look at some of The Source's features.

ON THE UPBEAT

Let's begin by tackling a common problem: suppose you've written a SourceMail message off-line with your word processor and saved it on a disk. You've made sure that the letter's been saved in plain old everyday ASCII (and not some compressed format that your word processor may be partial to). You've also checked to make sure your communications program has a way to send that ASCII file.

How do you get it from your disk to The Source?

The most straightforward way is simply to go to the SMail program and start transmitting. You could log on and, at the Command Level's -> prompt, enter MAIL S (for Mail Send). Then you'd fill in the "To:" and "Subject:" information, as always. After the "Enter text" prompt, you'd fire away by telling your communications program to start sending.

The Source neither knows nor cares that we're uploading rather than typing the letter in manually. That means it expects us to follow the same rules we would if we were typing it in manually.

That includes making sure that no lines begin with a period (remember, that's how all the "dot-commands" begin). Also there needs to be a carriage return on every line before the 78th character, just as if you were typing it in manually. There are a couple of ways to handle the carriage return requirement. First, check the manual that came with your communications program. It may be that your software has an option that "formats" the text as it's sent. If not, perhaps your word processor can format the text to 78-character lines when it saves it on a disk.

Ignore all this at your own risk. The Source is not amused at receiving lines that are too long. Some Source features just snip off the lines beyond the 78th character; others start sending you error messages even while you're transmitting. It gets ugly after that.

Also you can upload directly into your User File Directory where The Source has a couple of utilities specifically designed to keep up with your uploadings.

FILETRAN: THE UPLOADER'S FRIEND

In an effort to serve all the different kinds of communications programs we subscribers use, The Source has two kinds of uploading utilities for text files. First is FILETRAN.

To use it, go to the Command Level prompt and type:

****

-> FILETRAN <RETURN>

****

This will tell The Source to fetch the FILETRAN program and let you upload directly into your UFD, as if you had typed it in manually.

In a moment, The Source will display:

****

Enter number of lines to transfer between
acknowledgments:

****

This calls for a word or two of background. FILETRAN does *not* display your file as it's received. (If you're a jargon junkie, think of this as "half duplex.") The system is asking if you'd like to be sent messages along the way as it's receiving the file, a kind of status report. If you type 20 <RETURN> at the prompt, FILETRAN will report "20-lines transferred" after the first twenty are received, "40-lines transferred" after it gets the next batch, and so forth. Just a little extra service for your peace of mind.

After you type a number for the acknowledgments, the system says:

****

Enter file name:

****

It's asking for the name you want the file to have as it will appear in your UFD. You type in a name, say, NOTE <RETURN>. The Source says:

****

Begin your local procedures for file transfer
(Enter CTRL-P or BREAK when finished)

****

And here's where you tell the communications program to begin transmitting the ASCII file. You can use a bulk transmission here, if you'd like, or you can send it one line at a time.

Remember the carriage returns, though. FILETRAN is like most other features on The Source—it wants a carriage return on every line.

Note: FILETRAN can take up to 132 characters before it requires a carriage return. That's great, but . . . if you plan to use this file in SourceMail or POST, you'll have a problem if the lines are longer than 78 characters (the line limit for those features). We'd suggest you avoid possible confusion—just pretend that 78 characters is the maximum line length for text files throughout The Source.

If you were on-line right now and using FILETRAN, the system would not be displaying the file as it's received, but, as we noted, it would be giving you acknowledgments every twenty lines (or whatever number you requested back there). That continues until all the file has been transmitted.

Then you enter CONTROL P and the system would report something like

****

File transfer complete
154-total lines transferred

****

and return you to the -> prompt.

Once the file is in your UFD, you can manipulate it as you would any file. You can edit it with ED NOTE; you can type it out on your screen with TY NOTE, etc.

If the file were destined for your Sharefiles, you could enter PUTSHR NOTE. If it were heading for SourceMail or PARTICIPATE, you could go to the SMail or PARTI program and, after the "Enter text" prompt, you could type

****

.LOAD NOTE <RETURN>

****

(that is, dot-LOAD NOTE).

In fact, as you've learned in writing on-line, you can use the EDITOR program in most of The Source's message features. And part of the EDITOR is the LOAD command for moving files from your UFD. So the only new stuff here is the FILETRAN command that will let The Source receive directly from your disk.

WHEN FILETRAN WON'T DO THE JOB

A characteristic of FILETRAN, then, is that it doesn't "echo" your text back to your screen as it's receiving it. Instead, it sends those acknowledgments.

This is annoying for some subscribers; it's worse than that for others. Some communications programs are written in such a way that they need to "see" the echo coming back to them as they upload. For them, FILETRAN won't work.

That's why The Source has another program, RCV, that works in a similar way. To use it, go to the Command Level and type:

****

-> RCV <RETURN>

****

Like FILETRAN, the system will prompt you for the name of the file as it will appear in your UFD, and then say, "Begin your local procedures for file transfer."

Unlike FILETRAN, RCV prompts you for each line of the file, using a question mark as the prompt. It also displays the lines as it receives them. So, if you were to upload a short ASCII file, sending it one line at a time, your display might look something like this:

****

```
?    This is the first line of the file
?I'm calling NOTE. With RCV, The Source
?prompts me for each line of transmission
?using a "?". Of course, the question marks
?themselves are not saved in my file.
?(etc. more text, then)
?$$DONE <RETURN>
```

****

To close the file at the end of the transmission, RCV requires you to type $$DONE <RETURN> on a new line, as we've done in this example. That will return you to the -> prompt.

Of the two, we prefer FILETRAN, just because it's easier to send bulk transmission. However, RCV is the ticket if your communications software doesn't recognize flow control but can at least send one line of text at a time.

ON THE DOWN SIDE: DOWNLOADING

Capturing text files from the system for reading off-line doesn't require any special utilities from The Source—all the work's done on your end of the connection. It just depends on whether your communications program is capable of opening a buffer and capturing material, and then saving that information on a disk. If it is, all you have to do is go to your favorite Source features and open the buffer. After you're off-line, you can read your captured material by loading it into your word processor.

There's another kind of downloading that's popular with many subscribers: getting free public-domain BASIC programs from other Source members. It's easy and requires just a little bit of technical knowledge about your computer.

For instance, most microcomputers these days can run BASIC programs that are in ASCII form (though they often have a "compressed" BASIC as well that's usually a little faster).

It would seem reasonable, then, since ASCII is the "native tongue" of The Source, that you could capture a BASIC program on the system, downloading it as if it were an article or a letter, and save it to disk. Then, after you're off-line, you could call up your BASIC interpreter, load that ASCII file into memory, and run it.

Just so. That works for probably 95 percent of the computers that link up to The Source these days. All you may need to do is load the program file into your word processor to remove any non-BASIC lines that were captured along with the program (such as prompts and systems messages). Then be sure your word processor saves the edited version as an ASCII file, not a compressed version.

And, of course, you need to make sure the program you've downloaded from the system was written for your kind of computer. The BASIC language varies from one computer manufacturer to another.

If you'd like to experiment with this, watch the Sharefiles areas of The Source. Often messages on POST will speak of public-domain software available. You'll just need to open your program's buffer and read the shared file with the TYPSHR or the PUBFILE commands we've learned. (The POST message should tell you the command string you need.)

Also, of course, you might want to offer your own public-domain programs in your Sharefiles. You could upload them with FILETRAN or RCV and then put them in the public area with PUTSHR. The BASIC programs you upload need to be in ASCII form, of course, so you can transmit them just as if they were letters or articles. Incidentally, with BASIC programs you might want to take advantage of the fact that both FILETRAN and RCV can accept lines of up to 132 characters. That's so you can avoid chopping BASIC programming lines in two.

GRADUATE SCHOOL: BINARY FILES

Not all programs are BASIC. In fact, most of the quality software you buy these days is written in assembly language to take advantage of the computer's speed and graphics.

Most people who send and receive assembly-language programs by modem use that other method of uploading and downloading—the binary, or "error-free," transfer.

The Source has one of those utilities on-line. It's called KERMIT and was written a few years ago by people at Columbia University. If the KERMIT utility is part of your communications program, your manual should include instructions for using it. Then all you really need to know is how to have The Source call up its KERMIT utility. Easy enough—just type KERMIT <RETURN> at the -> prompt. People on The Source who use KERMIT usually work through Sharefiles. You can copy the shared file into your UFD with GETSHR, then summon KERMIT to handle transferring the file to your disk.

Another popular binary file transfer utility in use these days is one called Xmodem, a public-domain program written by Ward Christensen of Chicago. It's in use on a number of computer bulletin-board systems around the country. At present, The Source does not have Xmodem available. However, officials there say it may be added someday. Watch for the announcement.

Finally, some Source members upload and download binary programs in an ASCII form. (Smart things, these computers!)

They use a BASIC program that looks at a binary file and produces an ASCII representation of it that can be sent and received, just as if it were a text file. (This is often called an "ASCII HEX" file.) On the other end, the receiver downloads the ASCII file and runs it through the BASIC conversion program that turns it back into a binary file.

Your software package may have that kind of BASIC program already in it. (Look in your manual. It may be listed as something like CONVERT. BAS, or be referred to as ASCII HEX or ASCII-to-binary conversion.) If not, you may find a conversion program for your machine in the public domain. Ask around the system.

SUMMARY

This evening we covered two major areas of the system—Sharefiles and the wonderful world of file transfer.

We learned that you can add a public room on to your private User File Directory, where you can display files for other subscribers. You can also visit Sharefiles offered by others. These were the commands we learned:

TYPSHR sets up a sequence with which you can view a file in someone's Sharefiles area. You'll be prompted for the user ID of the file's owner and for the file's name.

GETSHR works like TYPSHR except that it copies the shared file into your UFD.

SHAREFILES opens your own SFILES area for business.

PUTSHR copies a file from your UFD to your SFILES area.

ASF "attaches" you to your Sharefiles area, that is, moves you from your UFD to SFILES.

ACL "moves" you back to your UFD from your SFILES.

PUBWARE is the name of the public software provided by The Source's special STCPUB account. Once copied, CO PUBWARE will gather the rest of the software needed to set up a menu-driven area in your Sharefiles.

PUBFILE (followed by a user ID) lets you visit the menu-driven SFILES area set up by a subscriber.

We also learned that uploading and downloading are a natural part of The Source and they depend on your equipment being able to play ball. The Source has two primary utilities to receive your uploading:

FILETRAN It doesn't echo the material back to your screen as it's received. It does, however, send you "acknowledgments" after a specified number of lines are received. You close the file with a CONTROL P or a BREAK key.

RCV does echo the received material back to your screen. It prompts you for each line with a question mark (?). You close an RCV'ed file by typing $$DONE <RETURN> on a new line.

The Source also has an error-free file transfer program called KERMIT. To use it, you need to have KERMIT as part of your communications program. For more information about KERMIT, contact Columbia University Center for Computing Activities, 7th Floor, Watson Laboratory, 612 West 115th Street, New York, NY 10025.

CHAPTER 13

Visiting the Library

We've come off the road from another series of complicated excursions into the system. There's no question about it—there's exciting potential in the features we've just seen: PARTICIPATE, SHareFiles, PUBWARE, and up- and downloading. Now that you've got the particulars, you'll master them with just a little old-fashioned practice. Also, as with anything new and complex, you should give yourself a little time for all those specifics to sink in.

The last time we needed a bit of a break, we headed off in new directions to see what we could discover in areas of The Source other than the Communication Services. That worked well, so let's do it again.

In this chapter we'll look at one of our favorite areas on The Source, the searchable databases. We've not hidden our enthusiasm for how well these electronic libraries work. When you've met some of them on the tours, you put them through their paces. Now, with a closer look at how they work, you'll be able to polish your searching skills. This also will give us a chance to point out some other good ones you might want to explore further.

In the next chapter we'll take another 90-degree turn and examine a different part of the system—its impressive array of business and investment services. The Source has developed quite a reputation for its fine financial utilities. This'll show you why.

After that, we'll come upon the finale:

First, you'll see how to bring together all the on-line tools you've acquired in order to customize The Source for yourself. In a sense, parts of every chapter in this book are contributing to that project.

Then we'll look ahead to what tomorrow might be like around The Source. Particular attention will be paid to an exciting new feature, the special-interest groups. You'll see that, while this is a new service, it incorporates some old friends you've already met in your travels.

This is all to say that, while we're coming down the home stretch, the best is not all behind us.

DOWN IN THE DATABASES

Based on first-hand experience, what have we learned so far about The Source's reference facilities?

For one thing, they are almost all searchable, usually by keywords, sometimes by date and other information as well. Most are quick, but can be even quicker if you know how to string commands together. It also helps to know something about how they use keywords.

Also, you've learned that the format varies from database to database. The prompts we used in searching the UPI database of news stories were not at all like those in the Member Directory or the airline schedule guide.

New subscribers often are put off by all this variety in the way the reference libraries work. Why can't they just pick one style for their databases and stick with it?

Actually, the format and style of a database usually depend on the kind of information it's offering. Some just need more searching specifications than others. It's the same thing with books. You would hope that Barbara Tuchman's latest historical study would include a good index for cross-reference. The same "utility" in a new novel would be ridiculous.

Despite differences in approach, the theory behind on-line searchable features is almost always the same, and not hard to grasp, particularly if we look at an overview of the databases and identify their major components.

DISSECTING A DATABASE

Just as a library depends on your knowing how to look up the location of the information you need, a database counts on you to be an active participant in the electronic search. After all, you, not the computer, will be calling the shots and making the tough decisions. But, then, you're so good that you make some of the decisions without even thinking about them.

For instance, when you visit one of the system's searchable features, you're actually standing at the threshold of a vast storehouse of *related* information. It could be a collection of software and hardware reviews or abstracts of recent business articles or today's and this week's stories from a major newspaper. In other words, you've already taken care of the first important step by choosing the right database for the job.

Usually, the first thing you see upon arrival is a list of categories describing what's available here. For instance, if you visit the system's Bylines service, a collection of nationally syndicated columns, you're greeted by a list something like this:

****

BYLINES on The Source

1 Viewpoints
2 The Arts
3 The Locker Room
4 Around the House
5 Confidentially. . .
6 Leisure
7 Health and Technology
8 Lifestyles
9 Search All Stories
<I>nstructions

Enter Item Number or <H>elp:

****

This is helpful, but it isn't an index (any more than it would be a bill of fare if a restaurant owner were to say that, generally speaking, he has appetizers, entrees, desserts, and beverages).

What the menu does is let you make another important decision by narrowing the options—"I think I feel like having a little commentary this evening. Take me to this 'Viewpoints' section." (And notice that here in Bylines, as in most of the system's databases, there's also an option that allows you to search all the entries, regardless of category.)

After you've made that selection, The Source is ready to get down to some serious searching, often with a prompt something like this:

****

<T>itle, <AU>thor,
<K>eyword, <D>ate, or <H>elp:

****

Nothing new here. We've seen prompts like this many times on the system. But let's think a minute about what this is really trying to tell us. This means that there are four ways to reach any one of the articles or abstracts in this section of the database. In this case, each article (or "record") has a title, an author, a date of publication, and keywords.

Other databases have many more than four such pathways. Remember the options in the Member Directory? They looked like this:

****

<ME>mber ID <NA>me <CI>ty <ST>ate
<CO>untry <O>ccupation <MI>icro <IN>terest
or <ALL>

****

In another kind of database, such as Microsearch, which houses thousands of computer-related articles and reviews, this first prompt is:

****

<K>eyword, <PU>blication, <TI>tle, <AU>thor,
<D>ate, <MI>cro, <O>perating System,
<MA>nufacturer, or <H>elp:

****

No matter what database you're using, it's at this first prompt that you set your search strategy in motion. For example, in the Bylines feature, you might specify WASHINGTON as your first keyword, and the database would search for a few seconds and then report something like, "71 stories found."

In most of The Source's databases, that report is followed by a second prompt, something like this:

```
***********************************************************************

     <N>arrow, <E>xpand,
     <R>ead, <SC>an, or <C>ancel:

***********************************************************************
```

The system is asking for directions. "Are seventy-one records too many for you to read—should I 'narrow' the group? Do you want more in the group—should I 'expand' it? Do you want to read all seventy-one of them now, or perhaps scan them and pick the ones that look interesting? Or should I just forget the whole thing?"

As you've seen on the tours, choosing either Narrow or Expand will cause the program to take you back to the first prompt to select another field—perhaps another keyword or a date of publication or an author, etc. The important thing to remember is that the database keeps the group of records it's found so far and then whittles it down further or adds to it following your next instruction.

You also have the "forget it" option. That's where the "cancel" fits in. It tells The Source, "Hey, forget that search we just performed." For instance, we found seventy-one records in Bylines by specifying WASHINGTON as a keyword. Suppose we narrowed that group by asking for just those whose AUthor was JACK ANDERSON and end up with twelve records. We then could change our mind and select the Cancel option. That would cause the database to ignore our JACK ANDERSON search and go back to the group of seventy-one. It would work the same way if we expanded the group. Cancel always makes the database pretend that the latest search never took place.

Incidentally, Cancel is a wonderful way to step back through a complex search and branch out in a new direction with the same basic group of records. Each Cancel rubs out a search strategy and backs up one step. Some databases you'll meet on the system don't specifically offer a Cancel option. However, the navigation command P (for Previous) usually will serve. Just enter P to backtrack through the search strategy.

Some of the system's databases have a few more options on the secondary prompt. For example, in the Microsearch service, the secondary level is actually two prompts. Right after your first search, you see:

```
****************************************************************

     <N>arrow, <E>xpand or RETURN for all:

****************************************************************
```

Narrow or Expand takes you back to the primary option prompt, but if you press the RETURN key (leaving the line blank), you see:

```
****************************************************************

     <R>ead, <SC>an, <ME>nu, <PR>int,
     <C>ancel:

****************************************************************
```

This version of the prompt has all three of our old friends—Read, SCan, and Cancel. In addition, there are two new ones.

MEnu? It's similar to SCan, except instead of giving you the first two or three lines of each record in the group, MEnu gives you only one line and a number with which to retrieve it, such as:

```
****************************************************************

     1 3/27 WASHINGTON MERRY-GO-ROUND (JACK ANDERSON)
     2 3/25 ANDERSON SPECIAL WEEKLY (JACK ANDERSON)
     3 3/26 WASHINGTON MERRY-GO-ROUND (JACK ANDERSON)

****************************************************************
```

PRint is similar to Read. However, the Read command, as you have already discovered on the system, displays the text with a "pause" on the screen every twenty lines or so, something like -More- or Enter RETURN to Continue. On the other hand, the PRint option is a continuous display, without such prompts. That's handy if you're downloading an article into your program's buffer, as we described in the last chapter. And, of course, here and everywhere else on the system, a CONTROL P will interrupt the display and return you to the preceding prompt.

LIGHT AT THE END OF THE TUNNEL

Looking at the databases in this light, we see something that should make all of them easier to use: despite differences in the way the various databases look, there really are only two basic kinds of prompts in use:

1. A *primary* prompt, at which you cut the records down to size or add some more to the pile.
2. A *secondary* prompt or prompts, at which you generally decide whether to hack at them some more (that is, return to the primary prompt), cancel the search and return to the primary prompt, or examine what you've found by reading or scanning.

So, no matter how the prompts are written on the various databases, there's no major difference in how they function.

But that's not to say there aren't other differences, ones that are subtler. For example, the way you can use keywords differs from one database to another. The same is true about the use of dates and time elements as a search specification.

Let's look generally at some of the differences in terms of keywords and dates. Then we'll apply what we've learned to some specific databases. After that, you should consider a little on-line practice on your own with the searchable features that strike your fancy.

THE KEY TO KEYWORDS

On our tours so far, we've used keywords in a simple, straightforward way. We just entered the word we were looking for, such as COMPUTER or APPLE. That means we really haven't been using keywords to their full potential. For instance, many of the The Source's reference features let you enter a phrase as well as a single word. On some of them, you can enter HARD DISK and the system will search its records for those that contain both of those words, side by side, as in HARD <space> DISK. It might not call your attention to entries that contained only HARD or only DISK.

Some, but not all, let you enter more than one keyword or phrase. They usually require that the words be separated by commas. You might be allowed to type in APPLE,IBM,TANDY and the system would find all the entries that contained at least one of those words.

Some give you an AND/OR option. With them, you might enter at the Keyword prompt something like: IBM AND SOFTWARE AND EDUCATION. You would mean, "Find all records that contain the words

IBM and SOFTWARE and EDUCATION. The words don't have to be side by side or even in that order, but pick only the entries that contain all three words." And on the same database, you might enter: TANDY OR COMMODORE and mean by that: "Find records that have the word TANDY or the word COMMODORE in them; they *don't* have to contain both words to be of interest to me."

Some Source databases want only complete words for keywords, such as COMPUTER. Others, however, will let you use parts of words, such as COMPUT. That's a nice feature. COMPUT would find entries that contain COMPUTER, COMPUTE, COMPUTING, COMPUTERIZE, COMPUTERS, COMPUTERIST, COMPUTATIONS, and others. In your electronic travels, you'll find other searchable features that will let you use partial keywords only if you telegraph your shots with an asterisk (*). You would enter COMPUT* for your keyword if you wanted to indicate "This is a partial entry."

Hmmm—it's beginning to look as if there are quite a few differences, huh? You may be wondering if this trip is really necessary. It is. Honest. Particularly when you consider that using the special features of a database will pay you back in connect time. The more you know about a database's local customs, the better you can make it work harder (and faster) for you. Certainly don't let the differences discourage you from at least exploring. We'll help.

In a minute we'll be looking at some major databases on The Source and giving you the specifics of how they handle keywords. Then later, if you're on-line and need a quick refresher course, it's easy to obtain. Just select the Keyword option from the primary prompt, and then enter the word HELP at the next prompt. In most cases, the system will give you a brief rundown on how that particular service uses keywords. If that doesn't work, most databases have an option for "Instructions," usually summoned by typing I <RETURN> at the primary prompt.

A few more things before we get into the database roundup.

First, about inputting. Many of The Source's databases let you save a little time with a kind of mini–command string. For instance, at a primary prompt such as this

```
*****************************************************************

   <K>eyword, <PU>blication, <TI>tle, <AU>thor,
   <D>ate, <MI>cro, <O>perating System,
   <MA>nufacturer, or <H>elp:

*****************************************************************
```

you *could* simply enter K for Keyword. However, usually you can enter K APPLE or K IBM. That'll cause The Source to skip the next prompt and get right to the searching. (On these databases, the command strings work on other fields, too, such as PU BYTE or AU PEYTON or MI KAYPRO, etc.)

And a word or two about how you select your keywords. Several pages ago we mentioned that a database is a collection of *related* information. That's an important point to remember when you're choosing your keywords. You'd be surprised how many people waste their good connect time and money by searching a database of computer-related magazine articles with the keyword of COMPUTER. Now, come on—wouldn't you expect the word COMPUTER to be in just about every record in that data storehouse?

So, Rule Number One—make sure you know the specialty of the database you're using and don't waste your time searching for the obvious, such as EDUCATION in an education-specific reference area.

And, while we're at it, Rule Number Two—when possible, plan your searches off-line and jot down some notes to refer to after you're connected. You now know that virtually all of the system's searchable features use the primary-secondary prompt system. So all you really need to know is what that primary prompt looks like, since it will give you the list of available fields that can be searched. If you knew that the primary prompt looked like this

```
***************************************************************

  <K>eyword, <PU>blication, <TI>tle, <AU>thor,
  <D>ate, <MI>cro, <O>perating System,
  <MA>nufacturer, or <H>elp:

***************************************************************
```

you could make some important decisions before you ever logged on. Suppose you were going to use this database to look for published reviews of specific programs for your local school system. Before you logged on, you might decide:

"First, I'm going to search the <MA>nufacturer field for APPLE. Then I'll <N>arrow the group down by searching the <O>perating System field for DOS 3.3. Then I'll <N>arrow further with a <K>eyword of EDUCATION. And finally, I'll look at the newest releases by <N>arrowing down the group to those reviews published since the first of the year, using the <D>ate field. . . ."

After you've made all those decisions, you could simply log on, visit the database you'd picked out, and follow the script you've written for yourself.

Which brings us to this curious Date field. . . .

KEEP ON-LINE TIME

At first blush, it appears The Source has no standardization when it comes to Date prompts. Newcomers complain that some searchable features want the information entered with numerals (2/11 or 2/11/85, etc.). Other databases want it a little more specific: Feb 2 1985 or 2 Feb 1985 or 2 Feb.

We hear grumbling and shifting in seats. But don't leave yet—there's a reason for this. It goes back to what we were saying about the structure of the searchable features: it differs from database to database because of the kind of information it's dealing with.

Some of the reference features, such as news wire services, keep a relatively small backlog of historical records—usually about seven days' worth of stories. Using them, it's simple to enter date information: "2/11-" would mean, "Give me everything you've got from Feb. 11 to the present" and "2/11-2/13" would mean everything between Feb. 11 and Feb. 13.

However, other databases keep several years' worth of abstracts or articles on hand. They usually want more specific references in the Date field, perhaps "11 Feb 1985-13 Feb 1985."

So how's a teletraveler to know which format to use?

There are three ways:

For immediate relief, read on. Date formats will be part of our upcoming summaries of specific features.

Later, if you're on-line and need help, just ask for it. Sometimes entering HELP at the Date option will tell you about formatting. If not, check the Instructions option.

Finally, there's often another option—List. Many of The Source's searchable features, particularly those with many records on-line, will add List to a prompt, such as:

```
******************************************************************

Enter date, <L>ist or <H>elp

******************************************************************
```

On these databases, if you select List, the system will display all the dates for which it has records, something like this:

```
*****************************************************************

  1 FEB 11, 1985
  2 FEB 12, 1985
  3 FEB 13, 1985

*****************************************************************
```

All you have to do then is indicate with the menu number which date you're interested in. (On most of the searchable features, once you've selected the List option, the system is expecting a menu number rather than a date at the next prompt. In this example, The Source wants you to enter 2 rather than FEB 12, 1985. One value of the List option is to see the date format for future reference.)

By the way, the Date field isn't the only one that may have a List option available. Some searchable reference features use special codes for publications or journals and refer you to a List option to get the specific codes. Others might want state abbreviations in a special style and offer a List to familiarize you with them.

Back to dates and times. Most of the search features let you specify a span of dates. Usually a hyphen (-) is the connector. For example, 2/11-2/13 (or perhaps 11 FEB 1985-13 FEB 1985) means records from Feb. 11 to Feb. 13. You usually can indicate "from this date to the present" in this style: 2/11- (with nothing after the hyphen). "Everything up until a certain date" could be -2/13 (a hyphen followed by a date).

Several databases will let you pick out dates, separating them by commas, such as 2/11,2/13,2/15.

On features that have new material added daily, such as the news databases and bulletin boards like POST, you usually can just tap RETURN at a Date prompt to mean "today."

GETTING SPECIFIC

You now have a good background for exploring nearly all the searchable features, so let's compile dossiers on the major ones, with marginal notes on any local customs we come across.

We'll start with some new ones we've promised, then back up to collect the same information on the databases we've seen in previous chapters.

Microsearch

As diverse as all of us are, there's one thing all The Source's visitors have in common: everyone's got a computer. We've found that people who use these marvelous machines generally have at least a passing interest in computer information: news of new ones, reviews, newly released software, etc.

A primary warehouse for computer information is Microsearch, a database of thousands of summaries of articles and reviews from some two hundred personal computer magazines and journals. Provided by Information, Inc., of Washington, D.C., the service also contains product literature from more than three thousand manufacturers.

Reaching It: Microsearch is accessible by following the Personal Computing branch of The Source's main menu. Easier, though, just enter MICROSEARCH at the -> prompt.

Main Menu: The Source is planning some alterations to this service, but the overview should look like this:

```
****************************************************************

MICROSEARCH (tm)

Copr. 1985 Information, Inc., Wash. D.C. All
rights reserved.

1 About MICROSEARCH (tm)
2 Magazine Article Summaries
3 Hardware & Software Summaries
4 Directory of Manufacturers

<I>nstructions

Enter Item Number or <H>elp:

****************************************************************
```

Primary Prompt: You saw it earlier in an example:

```
****************************************************************

<K>eyword, <PU>blication, <TI>tle, <AU>thor,
<D>ate, <MI>cro, <O>perating System,
<MA>nufacturer, or <H>elp:

****************************************************************
```

You can string together commands here, such as MA APPLE or O LDOS.

Keywords: You can enter more than one keyword if they're separated by commas (such as APPLE,IBM). No partial keywords are permitted, but you can use two-word phrases as long as there's a space between the two words (HARD <space> DISK).

Dates: Should be formatted as 3 FEB 1984-3 MAY 1984. You also can get a List of the publication dates on-line by entering L at the Date prompt.

Management Contents, Ltd.

This is a business library that contains abstracts of articles from leading business journals since 1982. It includes such big-league publications as *Business Week*, *Barron's*, *Advertising Age*, *Forbes*, *Harvard Review*, *Money*, *Economist*, and *The Banker*.

With CONTENTS, you can search out and read the abstracts on-line. Then, if you'd like, you can place an on-line order for a copy of the complete article for $10. The bill will be shipped with the order.

Reaching It: You can get to Management Contents by following the Business and Investing branch of the main menu, or by entering CONTENTS at the -> prompt.

Main Menu:

```
********************************************************************

MANAGEMENT CONTENTS, LTD.

1 Abstracts
2 List of Journals
3 Ordering Articles
4 Using Management Contents
Enter Item Number or <H>elp:

********************************************************************
```

A word about Option 2, "List of Journals." CONTENTS expects specific codes for the journals and publications. If you're a frequent user of CONTENTS, you might want to get a printout of the latest list for your files. Meanwhile, in the back of the book, we've placed the latest list available to us.

Primary Prompt: When you select Option 1, "Abstracts," you receive this prompt:

****

<K>eyword, <TI>tle, <AU>thor,
<J>ournal, <D>ate or <H>elp:

****

Keywords: Similar to the format on Microsearch. Multiple keywords can be entered, separated by commas. No partial keywords. Key phrases must have spaces between the words, such as COMPUTER LAW. If you're searching the title of an article, you might want to use a phrase with which you think it begins.

Dates: The format, like Microsearch's, is Day Month Year, as 2 FEB 1983.

United Media Features

This is Bylines, which we saw at the beginning of the chapter. It contains the complete text of syndicated columns on politics, movies, sports, health, technology, and more.

Reaching It: Go through the News, Weather and Sports option of the main menu, or, at the -> prompt, type BYLINES <RETURN>.

Main Menu: Upon arrival, you should see:

****

BYLINES on The Source

March 2, 1985

1 Viewpoints
2 The Arts
3 The Locker Room
4 Around the House
5 Confidentially. . .
6 Leisure
7 Health and Technology
8 Lifestyles
9 Search All Stories

<I>nstructions

Enter Item Number or <H>elp:

****

From this menu, if you choose any of the items 1 through 8, The Source will list the specific columns (such as Dr. Lamb, Jack Anderson, Miss Manners, etc.) and let you choose. However, if you'd rather choose by topic, select Option 9.

Primary Prompt: When searching, Bylines is driven from this prompt:

```
******************************************************************

<T>itle, <AU>thor,
<K>eyword, <D>ate, or <H>elp:

******************************************************************
```

Keywords: First of all, partial keywords *are* permitted in Bylines, but they must be signaled by an ending asterisk. For instance, entering ENGL* at the Keyword prompt would cause Bylines to search for ENGLand and ENGLish, among others. More than one keyword may be entered with commas separating them.

Dates: The format for dates is numeric—as 2/22-2/27. As usual, a trailing hyphen (2/25-) finds all entries from an earlier date to the present.

Washington Post

Monday through Friday, the *Washington Post* News Service updates its electronic edition. The selected stories have a decidedly political flavor, of course, but also include science and technology, and business and finance.

The system maintains a library of five back issues for searching.

Reaching It: Follow the News, Weather and Sports branch of the main menu or enter WPOST at the -> prompt.

Main Menu: It looks like this:

```
******************************************************************

THE WASHINGTON POST
2 March 1985

1 News Summary & Calendars
2 The Administration
3 Congress
```

4 Business & Finance
5 Science & Technology
6 The Courts & The Law
7 Politics
8 World & Nation
9 Editorials & Commentary
10 Search All Issues

<I>nstructions

Enter Item Number or <H>elp:

****

As with Bylines, the last item on this menu (Option 10) calls the search program into play.

Primary Prompt: On searching, the first prompt is:

****

<K>eyword, <D>ate, or <A>uthor:

****

Keywords: More than one can be entered if they are separated by commas. Also, partial keywords are possible but do *not* use an asterisk. For example, on this feature COMPUT would be used rather than COMPUT*.

Dates: WPOST looks for the Day and the Month, as 13 FEB. It will allow you to search more than one day as 11,12 APR and a range of dates as 14 DEC-17 DEC.

Scripps-Howard News Service

Another daily news service, Scripps-Howard is best for editorials and features from its member newspapers. It keeps today's reports and the previous six days' worth on-line all the time.

Reaching It: Again, through the News, Weather and Sports leg of the main menu, or by entering SHNS at the -> prompt.

Main Menu: It's short and sweet:

```
*****************************************************************

    Scripps-Howard News Service
    March 2, 1985

    1 Current Stories
    2 Search All Stories

    <I>nstructions

    Enter Item Number or <H>elp:

*****************************************************************
```

Either option takes you directly into the search program.

Primary Prompt:

```
*****************************************************************

    <K>eyword, <D>ate, or RETURN for all:

*****************************************************************
```

Keywords: For multiple keywords, separate them by commas. Yes, you can use partial keywords—no asterisk.

Dates: The format is numeric—2/11-2/15, etc.

ASSOCIATED PRESS

You've already used United Press International on The Source. The system also has the other major international wire service available, AP.

Associated Press Videotex keeps a seven-day database of world and national news, business, sports, and entertainment.

Reaching It: The News, Weather and Sports branch eventually will lead you to AP. To get into the express lane, enter AP at the -> prompt.

Main Menu: It looks like this:

****

THE ASSOCIATED PRESS
(C) ALL RIGHTS RESERVED
AP VIDEOTEX

1 News Brief
2 World and National News
3 Business and Finance
4 Sports and Entertainment
5 Search All Stories

<I>nstructions

Enter Item Number or <H>elp:

****

Primary Prompt:

****

<K>eyword, <D>ate, or RETURN for all:

****

Keywords: An important difference here. It recommends that multiple keywords be separated by *spaces*, not commas. For example, CRIME COMPUTER would find all stories that mention *both* "computer" and "crime," not necessarily side by side. It will ignore all those stories that mention only one of the words. However, you can also exercise an OR option. You can type in COMPUTER OR CRIME and the AP service will select stories that mention either "computer" or "crime" in the text. Partial keywords are possible, but they must have the asterisk—COMPUT*—to find those in the "computer" family.

Dates: Numeric—as in 3/15-3/21, etc.

The Career Network

We come here to demonstrate that not every database on The Source will conform to the system of prompts you've been studying in this chapter.

The Career Network is an excellent employment service provided by Computer Search International Corp. It is an on-line listing of candidates and jobs available in some forty industry groups. Resumes and job-opening announcements can be selected by state, region, salary range, etc.

This database uses the logic of the searches we've been talking about but not the displays we've been seeing. Instead, it spells out the alternatives in menus rather than brief prompts.

Reaching It: Go through the Business and Investing branch of the main menu, or enter EMPLOY at the –> prompt.

Main Menu:

****

THE CAREER NETWORK
COMPUTER SEARCH INTERNATIONAL CORPORATION (CSI)

ENTER ONE OF THE FOLLOWING OPTIONS:

OPTION	DESCRIPTION
1	EMPLOYMENT OPPORTUNITIES OR APPLICANTS AVAILABLE
2	EDITORIAL
3	TIPS ON RESUMÉ WRITING
4	TIPS ON INTERVIEWING
5	HOW DO I PUT MY RESUMÉ OR JOB LISTING ON "THE CAREER NETWORK"?
6	YOUR COMMENTS AND SUGGESTIONS

ENTER OPTION DESIRED (1-6) OR (Q)UIT:

****

After selecting the jobs or resumés options, you're prompted for the employment category code. Enter HELP to get a list. (Meanwhile, we've supplied the list in the On-line Survival Kit at the back of the book.)

Primary Prompt: After the program finds the entries in that job category, you receive this prompt:

****

AT THIS POINT, YOU MAY

1—PRINT THESE ENTRIES,
2—SELECT FURTHER FROM THIS DATA BASE BY SPECIFYING GEOGRAPHIC LOCATION, SALARY RANGE DESIRED, DATE OF ENTRY, ACCOUNT, SIC, FUNCTION, OR DEGREE CODES, OR WITH A KEYWORD SEARCH,
3—LOCATE AND PRINT AN ENTRY BY REFERENCE NUMBER, OR
4—RETURN TO SELECTION OF ANOTHER CATEGORY.

ENTER 1-4 OR (Q)UIT:

****

Selecting Option 2 gives you this prompt to work with:

****

YOU MAY SELECT ENTRIES FROM THIS DATA BASE BY

1—SPECIFYING SALARY RANGE DESIRED,
2—SPECIFYING GEOGRAPHIC LOCATION DESIRED,
3—PERFORMING A KEYWORD SEARCH,
4—SELECTING BY DATE OF ENTRY,
5—SELECTING BY ACCOUNT, OR
6—SELECTING BY SIC, FUNCTION, OR DEGREE CODES.

ENTER 1-6 OR (Q)UIT:

****

Keywords: The program lets you search in these fields: job title, type or size of business, product, education, or job description, experience, and comments. You may enter a single word. Multiple keywords also are permitted, but they must be connected with an AND or OR, such as NEWSPAPER OR JOURNALISM, SALES AND PROGRAMMING, etc.

Dates: They're entered in a different way here. The database will find entries made on or after a specified date. The date has to be entered numerically, without slashes or hyphen. For instance, June 1, 1985, would be 060185.

A–Z Online Travel Service

This service lists some 23,000 hotels around the world and can be searched for cities, rates, location, credit cards, and business accommodations and services.

Reaching It: Through the Travel Services menu, or by typing in AZ at the –> prompt.

Main Menu:

****

A-Z ONLINE TRAVEL SERVICE

1 A-Z Hotel Directory
2 About A-Z Travel Services

<I>nstructions

Enter item number or <H>elp:

****

Primary Prompt: After specifying a city, with an optional state or country, the database makes an initial search and then displays:

****

<N>arrow, <E>xpand, <REC>ap, <R>ead,
<SC>an, <PR>int or <C>ancel:

****

(Incidentally, the RECap option redisplays the findings, in case they've scrolled off your screen.)

Upon selecting Narrow or Expand, your next prompt is:

****

Narrow by <NA>me, <RA>tes, or <K>eywords:

****

Keywords: You can enter up to six, separated by either a space or a comma. Partial keywords are permitted and no asterisk is required.

BACKTRACKING

Now let's replow some old ground, looking at searchable features we visited in earlier chapters and making note of how they handle keywords and dates.

United Press International

This was one of the first reference features we visited. We reached it through the News, Weather and Sports branch of the main menu and by entering UPI at the Command Level prompt.

Keywords: UPI can accept up to three keywords, but they must be connected by either an AND or an OR, such as COMPUTERS AND SALES AND APPLE, CHESS OR POKER OR BRIDGE, etc. In this feature, a search cannot have both an AND and an OR in it. Partial keywords are permitted, as long as they have the trailing asterisk, such as TELECOMMUNIC*.

Dates: The standard numeric format—6/1-6/7.

The POST Bulletin Boards

We've visited POST several times and already used its search features extensively.

Keywords: It's important to remember that a keyword search on POST checks only the header material (the subject lines and keywords), not the text of the message itself. Multiple keywords can be linked with a comma. However, because of the nature of the service, we've found better results by using one-word searches and then narrowing or expanding the group with the subsequent prompt. Partial keywords are permitted; no asterisk is required.

Member Directory

This is the nifty users' log we visited early on to find out how to reach other subscribers to The Source.

As we've seen, MEMBERS lets you have keyword searches on eight fields: member ID, name, city, state, country, occupation, micro, interest (or, of course, all fields).

Keywords: MEMBERS generally is best searched using single keywords. However, you can use partial words and no asterisk is required. For instance, JOURN would find JOURNALIST, JOURNALISM, JOURNAL-WRITING, etc.

Mobil Hotel and Restaurant Guides

These are the guides we used in our tour of the travel, weather, and shopping features.

Keywords: First, as you recall from the tour, these databases call this Find rather than Keyword, but the operation is the same. They will accept one keyword at a time, or partial keyword, no asterisk required.

Note, too, that instead of Cancel, they use Backup—it's the same function.

CHAPTER 14

Taking Care of Business on The Source

Moments after The Source was born, some people predicted that its destiny lay, in part, in providing instant information to those who move in the world of business and high finance.

The company has continued to add to its stable of definitive business services. The information has become more diverse, and so has the audience. You don't have to travel in the fast lanes of commerce, however, to use the wealth of material available and to appreciate how quickly you can pinpoint it. If you do require computer tools for the "serious investor," you'll find enough utilities here, we think, to turn even your broker's head.

So in this chapter the tour bus goes "uptown" to look at some of the system's business and economic features. For instance, you'll see how to:

- Find today's stock market prices and other information on any security on any major exchange.
- Buy stocks on-line through a new high-tech investment service.
- Maintain your own portfolio in your User File Directory so that one quick command gives you the rundown on all the stocks you're interested in.
- Use the kind of analytical and projection programs that once were available only to a few.
- Get your hands on top-flight industry reports that you used to have to depend on your broker for.

• Read up-to-the-minute business news from all over.

• Learn about the "freebies"—inducements that make some stocks more attractive than others.

Even if you're not an investor, some of these areas can be an education. By setting up an imaginary portfolio, you can use the utilities to determine just how sharp your business instincts are.

Some of these are "value-added" services—that is, you pay extra to use them. However, our on-line tour will sidestep those areas, and instead we'll talk about them when we're off-line again. After hearing what the services have to offer, you can decide whether to use them.

One other thing before we log on: have a pencil and a notepad handy so you can scribble down a few things. That's the only extra equipment you'll need on this tour.

HITTING THE ROAD

Log on, turn off CHAT, and call up the main menu as we always do. This will show us the part of the system to which we're traveling:

```
*****************************************************************

MENU OF SERVICES <MENU>

1 Today From The Source <TODAY>
2 News, Weather and Sports <NEWS>
3 Business and Investing <BUSINESS>
4 Communication Services <COMM>
5 Personal Computing <PC>
6 Travel Services <TRAVEL>
7 Shopping, Games and Leisure <HOME>
8 Member Information <INFO> **FREE**

Enter item number, <H>elp or <Q>uit:

*****************************************************************
```

Note that the menu says if you're going directly to the Business and Investing area from the –> prompt, you can type BUSINESS <RETURN>. But since we're only a number away, let's press 3 <RETURN> to see:

****

BUSINESS AND INVESTING <BUSINESS>

1 STC/SSI Investor Services <INVEST>
2 Business Update <BIZDATE>
3 Investment Data and Analysis
4 Financial Market Reports and News
5 General Business Reference
6 Employment Services <EMPLOY>
7 Business Bulletin Board
8 Financial Services Index <BIZDEX>
9 Tax Information <TAX>

Enter item number, <H>elp or <Q>uit:

****

This gives you some idea of the scope of business services and we're not even at the end of the menu road yet. Choose Option 4 here to see:

****

FINANCIAL MARKET REPORTS AND NEWS

1 Financial Markets <UNISTOX>
2 Stock Prices from AP <STOCKCHECK>
3 Commodity World News <CWN>
4 Donoghue Moneyletter <DONOGHUE>

Enter item number, <H>elp or <Q>uit:

****

The tour bus is nearing its stop, but before we move on, take a close look at the current menu. These are some of the services we'll be talking about after we get off-line.

Now choose Option 2 which will take us to Stockcheck.

****

STOCKCHECK SERVICE

1 Latest AP Stocks/Bonds Quotations (c) copyright The Associated Press, 1984

2 Stocks/Bonds Symbol Lookup Table

<I>nstructions

Enter item number or <H>elp:

****

YOUR ON-LINE TICKER TAPE

Option 1 on this menu takes you to an area where you can get stock prices from the Associated Press. This is not an instant stock-reporting service. If you use this service while the markets are still open, there is generally a twenty-minute delay in the quotes, so "latest" in this case doesn't mean up to the minute. (If, however, you need undelayed quotes, stay tuned. Later we'll tell you about features in which "instant" means INSTANT.)

To use this service you must know the standard symbols for the stock or securities you're looking for. Finding them is what Option 2 is for, so let's go there first. Type 2 <RETURN> to see:

****

Enter Company Name or Press Return for Previous Menu

****

For this exercise, we're going to find symbols for three companies—International Business Machines, Tandy Corp., and Apple Computer.

This program will find a symbol even if you don't know the complete name of the company. All you really need to know are the first two letters of the first word in the company name. But, of course, the more of the name you know, the faster you can find the symbol you need.

For example, if you were to type only INTERNATIONAL <RETURN> at the prompt, in an attempt to find the symbol for International Business Machines, you'd get a list of every public-stock corporation with International as the first word in its name. There's a bunch of them. So instead, type in the full name, INTERNATIONAL BUSINESS MACHINES, and you'll see:

```
*****************************************************************

Company Name                                         Symbol  Exch
--------------------                                 ------  ----
INTERNATIONAL BUSINESS
  MACHINES CORP .................................... IBM     NYSE
Enter Company Name or Press Return for Previous Menu

*****************************************************************
```

No fair. You probably already guessed that one.

Next, we want to find the symbol for Tandy Corp. The system has returned you to the place where you can input another company. This time, type TANDY <RETURN> and you'll see something like this:

```
*****************************************************************

Company Name                                         Symbol  Exch
--------------------                                 ------  ----
TANDY BRANDS INC.....................................TAB     AMEX
TANDY CORP. .........................................TAN     NYSE
TANDY CP 10% DUE 1994................................TAN94   NYBD
TANDY CP 10% DUE 2004................................TAN91   NYBD
TANDYCRAFTS INC......................................TAN     NYSE
Enter Company Name or Press Return for Previous Menu

*****************************************************************
```

Two of these securities, called Tandy CP, are bonds on the New York Bond Market. The other three are stocks, two of which are on the New York Stock Exchange. The one we're interested in is Tandy Corp. Its symbol is TAN. On your notepad, write that symbol under IBM and let's find one more. Type APPLE <RETURN> and see:

```
************************************************************

Company Name                                Symbol  Exch
--------------------                        ------  -------
APPLE COMPUTER............................AAPL    OTC

Enter Company Name or Press Return for Previous Menu

************************************************************
```

The "OTC" tells you this is an over-the-counter stock.

Okay. Now that we have our three symbols—IBM, TAN, and AAPL—we're ready to find their status in today's market. As the prompt says, press RETURN and we'll go back to the previous menu:

```
************************************************************

STOCKCHECK SERVICE

1 Latest AP Stocks/Bonds Quotations (c) copyright The Associated
  Press, 1984

2 Stocks/Bonds Symbol Lookup Table

<I>nstructions

Enter item number or <H>elp:

************************************************************
```

Choose Option 1. You'll see this prompt:

```
************************************************************

Enter Stock/Bond Symbols or Press Return for Previous Menu

************************************************************
```

Now type in the first symbol on your list—IBM <RETURN>—and you should see a line of stock information, something like this:

****

Date	Time	Symbol	P-E	Sales (hds)	High	Low	Close	Chg.
03/04	18:15	IBM	13	8489	137⅞	136¾	137¼	+⅛

Enter Stock/Bond Symbols or Press Return for Previous Menu

****

Now do the same thing with TAN <RETURN>, then AAPL <RETURN>.

Seeing the reports that way is fine, but there's a better way. You don't need to enter them one at a time. You can list more than one symbol on the same line, with a space between each symbol. Try it:

****

Enter Stock/Bond Symbols or Press Return for Previous Menu
IBM TAN AAPL <RETURN>

Date	Time	Symbol	P-E	Sales (hds)	High	Low	Close	Chg.
03/04	18:15	IBM	13	8489	137⅞	136¾	137¼	+⅛
03/04	18:15	TAN	13	5916	32¾	31¾	32⅛	−½

Date	Time	Symbol	Sales	High	Low	Close	Chg.
03/04	17:36	AAPL	4804	20	19¾	20	+¼

Enter Stock/Bond Symbols or Press Return for Previous Menu

****

ELECTRONIC PORTFOLIO

You can also keep a "portfolio file" on-line in which you can store the symbols for specific stocks and get quotes automatically without having to type in all of the symbols each time. It's sort of like a financial command (or CO) file.

To do that, let's go back to the -> prompt. Press RETURN at the prompt on your screen—that takes you back to the main menu. Then type Q <RETURN> as always.

Creating our portfolio is the same as writing any other text file on-line; at the -> prompt, you need to type ENTER followed by the name of the portfolio file you want to create. You can call it nearly anything. Let's call ours MYSTOX:

****

-> ENTER MYSTOX

****

Now type in each symbol, one per line, using the symbols from our excursion. Your file will look like this:

****

IBM <RETURN>
TAN <RETURN>
AAPL <RETURN>
<RETURN>

****

The last <RETURN> on a line by itself closed the file, of course, and brought you back to the Command Level prompt.

You needn't worry if other Source subscribers have files named exactly the same. They'll not be able to look at the files in your private UFD.

Now, from the Command Level prompt, we're going to take the shortcut back to the stock program. At the -> prompt, type this command string (with a space between the "k" and the "1").

****

-> STOCKCHECK 1 <RETURN>

****

This takes you right to Option 1 on the Stockcheck Service, skipping all the menus you used to get there earlier. The next thing you see will be:

****

Enter Stock/Bond Symbols or Press Return for Previous Menu

****

Now type in MYSTOX, in parentheses, like this:

****

(MYSTOX) <RETURN>

****

The table of information about the three stocks will be displayed again.

Now, *that's* more like it.

After you've looked at it again, press RETURN to get back to the Stockcheck Service menu and then Q <RETURN> and return to Command Level.

We're going to log off now, but in the interest of saving storage charges, delete the temporary portfolio—DEL MYSTOX <RETURN>. Later you can come back and look up other stocks if you'd like to build a permanent portfolio. In fact, you can build several, if you'd like, giving them each different names. One could contain the symbols for stocks you own, another for stocks you'd *like* to own or have considered buying. Then, when you're using Stockcheck, just put in parentheses the name of the one you want to check.

Log off now and we'll tell you some other things you can do with an on-line portfolio.

During that brief tour, we pointed out the menu headings for some of the other powerful securities programs and business-information features. We can't cover all business services here and, in fact, there's a good chance some have been added since we passed that way. Still, we can hit the highlights.

For instance, you can use on-line portfolios we created in several important programs on The Source.

SOURCE/SPEAR SECURITIES: TRADING STOCKS ON-LINE

Ever since the dawn of computer-based information services, digital gurus have predicted that someday people would buy and sell stocks on-line from their offices or homes. In 1984 The Source and Spear Securities, Inc., brought the dream to life.

Option 1 on the Business and Investing menu takes you to an area of The Source in which you can trade securities through Spear, a Los Angeles-based brokerage firm that specializes in on-line securities trading at discount commission rates.

You need to have established an account with the company in order to trade. However, you don't have to be a Spear customer to window-shop and to use other powerful programs in that area.

The investor services area can be reached through the main menu or by typing INVEST <RETURN> at the -> prompt. The introductory menu looks like this:

****

STC/SSI INVESTOR SERVICES <INVEST>
THE SOURCE/SPEAR SECURITIES
TRADING AND PORTFOLIO MANAGEMENT

1 Quotes <INVEST 1>
2 Portfolio Manager <INVEST 2>
3 Spear Securities Trading Online <INVEST 3>
4 About Trading and Quotes <INVEST 4>
5 Financial Markets and Analysis
6 Business and Investment News

****

Option 1 of this menu leads you to a program that works very much like the Stockcheck Service. In fact, you could have used the list of stocks we created in the file called MYSTOX by typing in the file name at the "QUOTES" prompt to receive the same information you received in Stockcheck.

But there's a major difference here—time. If you are a serious investor, you know what a difference twenty minutes can make in the market. So, for an additional monthly fee, you can get "real-time" quotes in this area, as opposed to the twenty-minute delay on Stockcheck. Without subscribing, you get a twenty-minute delay here as well as in Stockcheck. To learn more about real-time quotes or to begin the process of signing up for the service, call The Source at 800-336-3330 or send SourceMail to TCA068.

One of the most impressive features of INVEST is "The Portfolio Manager." In this area, you can create special portfolio files with names and passwords for added protection.

In these files, you can add such information as the number of shares of a given security or stock you have purchased, when you purchased it, what price you paid for it, etc. Then, in a subsection of the service, you can get a composite list of unrealized gains and losses in the portfolios, a schedule of realized gains and losses, a list of holdings by tax lots with market values and other important information about your stock holdings. In short, many of the things you pay a broker's commission for.

You need not be a Spear customer to create a portfolio file or use the Portfolio Manager. You need not even purchase the stock—they can be imaginary "what if" portfolios. The only additional charge for this service is the extra storage costs for the files in your UFD.

If you *are* a Spear on-line customer, your portfolio automatically will be kept up-to-date for you by the system. Every trade you make with the company will be added to the portfolio by the brokerage firm. Also, you can specify market or limit orders, day or good-until-canceled orders, stop orders, cash or margin purchases, or short sales. Your market orders for up to one thousand shares of nearly four thousand frequently traded stocks are executed instantly during regular exchange hours at the best available price.

Options 3 and 4 on the Investor Services menu give explanations of the Spear service and how to get more information about Spear accounts, what it takes to be a Spear customer, and how transactions are handled.

Feel free to explore this area. Don't worry that you'll get lost and confused and end up buying a thousand shares of United Buggy Whips, Ltd., at $80 a share. You can't purchase stock unless you first fulfill the requirements for becoming a customer.

MEDIA GENERAL'S STOCKVUE

Stock analysts know how to take the numbers from the financial pages and divine The Big Picture, to tell us what's really happening and what's likely to happen in various markets. The Source's Stockvue program from Media General puts that kind of power in your hands, too.

You may remember seeing an option called "Historical Stock Analysis <STOCKVUE>" on one of the menus we passed. You can reach it by following the "Investment and Data Analysis" branch of the Business and Investing menu, or by simply typing STOCKVUE <RETURN> at the -> prompt. The Stockvue menu looks like this:

```
*****************************************************************

MEDIA GENERAL
STOCKVUE

1 OVERVIEW
2 INSTRUCTIONS
3 DESCRIPTION OF SCREENS
4 DATA ITEM DEFINITIONS
5 INDUSTRY GROUPS NUMBERS LIST
6 CREATING PORTFOLIO FILES
7 ** STOCK ANALYSIS AND TABLES **

Enter item number or HELP:

*****************************************************************
```

The first six of these seven areas may be accessed without charges beyond the usual connect-time costs. The seventh item—the area where the analyses take place—is a surcharged feature. (To find the current surcharge for the analyses, type RATES <RETURN> at the -> prompt.)

The database provides some 55 items of information on 3,100-plus stocks from the New York, American, and over-the-counter exchanges. All information is updated weekly.

The information is delivered to you in twelve groupings called "screens." You can think of a screen as a computer program that gets raw data on a stock or group of stocks, performs computations, and delivers the findings to your computer screen in a way that makes it easy to read and interpret.

If you choose Option 3 from the Stockvue menu, you'll see a description of the various screens of information available. A list of them is also included in the On-line Survival Kit at the end of this book. You should study this information before entering the surcharged Option 7 of Stockvue. Knowing what you want before going in will save you money.

Unless you're a financial wizard, some of the terms used in these analyses might confuse you. Fortunately, Option 4 on the Stockvue menu can help: it causes the program to list definitions. For example, you'll find that screen 6 displays several statistics relating to the market value of a security in relation to the total market and includes "up beta and down beta." Terms like that stump many of us, but the definitions in Option 4 would tell you:

"BETA UP/DOWN—Beta is a measure of how a stock tends to vary in price relative to the market. The Media General up betas are based on the latest eight upswings of 5 percent or more of the market. A beta of 1.00 means the stock has moved evenly with the market over the measured intervals. A beta of 1.50 means the stock has moved up 50 percent more than the market and a beta of .90 means that it has moved up only 90 percent as much as the market. Down betas are calculated over the last eight downswings of 5 percent or more in the market. And the interpretation would be the same as that for up betas, except of course, the measure is a relative onc on the downside."

Most of the terms used in the screens have definitions available under Option 4.

Now, about that portfolio file you created during the last on-line tour: you can use the same portfolio file in Stockvue. You can run not only single stocks, but a series of securities, through the analyses. (Sometimes you're better off performing an analysis on a group of stocks, since Stockvue, in many instances, will rank them during the computations.)

For example, we could have put MYSTOX through an analysis. Choosing Option 7, we would have seen this message:

```
*****************************************************************

WELCOME TO MEDIA GENERAL STOCKVUE
**YOU ARE NOW IN A SOURCE*PLUS SERVICE**
TYPE "HELP" or "QUIT" AT ANY TIME.
ENTER STOCK SYMBOLS, GROUP NUMBERS, OR FILE
NAMES
(TYPE "SYMB" TO FIND SYMBOLS OR "HELP" FOR ASSIS-
TANCE.)
!

*****************************************************************
```

The exclamation mark is your prompt for this service—you can enter the stock, series of stocks, portfolio name, or industry group you're analyzing. If you're entering the stocks one at a time, type one symbol after the exclamation mark and press RETURN. Remember, one stock per line. When you're finished, press RETURN on a new line.

You can analyze an entire group of stocks within a business category by entering an industry group number after the exclamation mark and pressing RETURN twice. A list of these industry group codes and the businesses they represent can be found under Option 5 of the Stockvue program or in the On-line Survival Kit at the back of this book.

However, the most efficient way of searching for stock information is to use a portfolio file you've already created. You could type (MYSTOX) <RETURN> at the exclamation mark, then press RETURN again when the next exclamation mark appears. The system would read the list of stocks from the MYSTOX file. Next, you'd see:

****

PICK A SCREEN (1-12)
TYPE "HELP" FOR A LIST,
OR "STOP" TO PUT IN NEW STOCKS.

****

If you refer to the list at the back of the book, you'll see that screen 2 divides the stocks in the portfolio by industry group, shows the stock name, symbol, exchange, current price, and the percent of change during the most recent week for each stock. This screen first asks you if you want to rank the stocks by percent change. Then you're asked if you want them listed in the rank from highest to lowest or lowest to highest.

At the end of the screen of information, you have the opportunity to pick another screen without having to type in the MYSTOX file name again.

To exit from the stock analyses section of Stockvue, type STOP at the end of a screen listing to take you back to the exclamation mark. Then type STOP again to take you back to the main Stockvue menu. From there, Q <RETURN> will take you back to the –> prompt.

It's clear that Stockvue puts a considerable amount of analytical power in the hands of telecommunicators. And while it is a surcharged service, careful preparation before you visit means you can be in and out in a matter of minutes.

UNISTOX: AN INUNDATION OF INFORMATION

On the *Tonight Show,* when Johnny Carson reads a news item or quotes from a new book, Ed McMahon often sets him up for a traditional routine, by saying something like, "Isn't it amazing that *everything* you'd ever want to know about that can be found right there in that list?" To which Carson always says, "No, no (characterization deleted)—there's more . . ."

Well, the old chestnut would have a new punchline if Carson had ever used Unistox. Everything he ever wanted to know about anything of value traded anywhere in the world seems to be on this list.

Unistox is built around more than two hundred reports United Press

International issues each day about securities of all kinds. Using it requires a little preparation on your part. Each of these reports is labeled with a three-digit number and each report appears approximately at the same time each day.

If you choose No. 1 from the Financial Market Reports and News menu or type UNISTOX <RETURN> at the command prompt, you'll see the following:

****

. . .UPI UNISTOX. . .ARE ON-LINE!
TYPE "HELP" OR "QUIT" AT ANY TIME. . .

WHICH REPORT?

****

Hmmm. See what we mean by preparation? The program is asking you to give the three-digit code for the report you want to see.

You should always begin your exploration of Unistox by studying the reports. You can get the latest list of major reports by typing INFO REPORTS <RETURN> at the –> prompt. The list begins like this:

****

DATA REPORTS IN ALPHABETICAL ORDER

4:23 PM	050	AMEX 4PM MOST ACTIVES
4:16 PM	148	AMEX 4PM STOCK & BOND PRICES
7:40 PM	035	AMEX CLOSING 50-SELECTED LIST

****

That means if you type 050 <RETURN> at the "WHICH REPORT?" prompt, you'll get a list of the American Stock Exchange's most-active stocks as of 4:00 P.M., which is put on the UPI wire at 4:23 P.M. (If you seek the information prior to 4:23 P.M. Eastern Time on any weekday, you're likely to get the previous day's report.)

Timing often is everything when you're using Unistox. That's why you can also get a list of reports in order of their appearance during the day by typing INFO TIMES <RETURN> at the Command Level prompt. Then you'll see the list of reports as they are sent out by UPI from early in the morning till late at night.

Once you know what Unistox offers, the numbers of the reports and the times in the day they are offered, you're set to get the information you need.

Much of this information appears in the form of lists, so sometimes you'll want to search with a keyword. After you enter the number of the report, the Unistox program allows you to enter keywords, one per line, to see selected information rather than the entire report. For example, the lumber futures report might contain a list of seven months, but you might be interested only in the price of lumber contracts for September delivery. Thus, you would enter "Sep" for a keyword, press RETURN twice and only the September price from the report would be listed.

To see the entire report at the keyword prompt, simply press RETURN, leaving the line blank. Since many of the reports are in text form, you often will want to see the entire text.

OTHER BUSINESS SERVICES

In covering databases in the last chapter, we slipped in three business-related services—The Career Network (EMPLOY), MICROSEARCH, and Management Contents (CONTENTS). These are important business tools, and there are more along the same lines.

Business Update <BIDZDATE>

Business Update is available from the entry menu or by typing BIZDATE <RETURN> at the Command Level prompt. The Source newsroom assembles this daily summary, and it may be just what you're looking for if you want a quick look at the day's business news. With it, you can get five or six of the day's top business stories, then go directly to Stockcheck or the commodities index.

Leading market indicators are accessible from BIZDATE. Another option gives you a line or two about each of the other items on the menu for those who want an even quicker look at what's hot.

Commodity World News <CWN>

If your interests lie in the wild world of commodity trading, Commodity World News has the latest quotes and news on the fifty-nine most active commodities. Price quotes are available through a program similar to Stockcheck. You may enter a commodity symbol, commodity name, or a file name in which you've listed the commodities in which you are interested.

Also, you can search using keywords and date limiters in much the same way as you search other news databases. You can access Commodity World News by typing CWN <RETURN> at the Command Level prompt.

Information on Demand, Inc. <IOD>

As a Source subscriber, you have a research department at your disposal. Information on Demand maintains a staff of researchers who have access to hundreds of computer databases, government agencies, trade associations, and major research library collections throughout the world. If you need research assistance, Information on Demand can do it for a fee. Some of the searches Information on Demand has done in the past include:

Market opportunities for alternate energy sources.
Executive compensation packages.
Safety regulations and plant procedures for workers with hazardous materials.
New technologies in genetic engineering.
Financial data on companies in the transportation industry.
Applications of videodisc technology.
International trade statistics for precious gems.

In addition to research and information, Information on Demand is a commercial document delivery service. If you need a journal article, technical report, conference paper, or nearly any other published material, the service can get you a copy, often overnight.

The service generally charges $60 an hour for its research time, with a two-hour minimum. To see more about Information on Demand, type HELP IOD at the Command Level prompt. To enter the Information on Demand area of The Source, type IOD.

There you can complete the questionnaire to request a search or send Sourcemail to TCB352. IOD will reply to your inquiry within twenty-four hours. You will be billed directly by Information on Demand for the services the company provides.

Donoghue Moneyletter <DONOGHUE>

Bill Donoghue's money newsletter is a menu-driven on-line magazine about money and investments. The newsletter provides information on such topics as the money market, mutual funds, all types of securities, banking, tax regulations, and much more. Each article is dated at the menu level to alert you to when the article was posted.

To access the newsletter, type DONOGHUE at the Command Level prompt.

<INVESTEXT>

The Investext database, one of the newer business services The Source has to offer, features the latest analyses, financial data, and market forecasts for more than three thousand companies. The reports are produced by two hundred analysts at some of the world's leading investment banking and research firms.

You can enter the Investext area by typing INVESTEXT <RETURN> at the Command Level. The actual reports are available in a searchable database, which can be accessed in various ways through the first four selections on the Investext menu.

This is a value-added service and to find out the current charges for obtaining reports, choose Option 6, "Value Added Charges," from the Investext menu.

Selection 8 from the menu, "Using INVESTEXT," gives you ten screens of information on how to use the service. These help files are not surcharged.

MODEL1

This value-added service is a sophisticated on-line financial modeling program. With MODEL1, you can create applications such as profitability analysis, cash flow, budgeting, product planning, project evaluation, net present value, internal rate of return, or an investment study.

Before you can work with MODEL1, you must have a user's manual. Type HELP SYSDOC at the Command Level for information on how to order the manual.

If you think you might be interested, you can explore MODEL1 on-line before you decide whether to order the manual. HELP MODEL1 typed at the Command Level prompt will give you a brief overview of the service, a list of some of the commands, and a description of what a MODEL1 program might look like once it is completed.

If you type HELP PROFIT at the Command Level, you'll see a typical profit-and-loss statement you can create with MODEL1. Only the program itself is surcharged, not HELP MODEL1 or HELP PROFIT. To find the latest surcharge for MODEL1, type RATES at the Command Level.

Shareholder Freebies and Discounts <FREEBIES>

If you're a stockholder, you probably know that many companies provide product discounts and benefits to their investors. This information often isn't well-publicized, however.

This service is a listing of those companies offering inducements to purchase stock. It is published only in electronic form and asks for feedback from users in building the data.

For the investor, this database can be vital when considering investments since discounts and inducements to purchase stock can amount to a considerable sum of money.

IRS Information <IRS>

If you type IRS at the Command Level prompt or choose number 9 from the Business and Investing menu, you'll see a tax-information service brought to you by The Source in cooperation with the Internal Revenue Service.

This database includes a current feature prepared by the IRS and a complete list of free tax publications which you may order on-line.

Tax Notes Today <TNT>

This feature by Tax Analysts, Inc., is a listing of current and past articles dealing with the latest government action that may have an impact on taxes as well as action taken by the IRS concerning the tax laws. This database is particularly important for tax attorneys and accountants who must keep up with current IRS rules and regulations.

WRAPPING UP

We've shown in this chapter how you can create a text file in your User File Directory to act as a sort of command file to get you the information you need quickly and efficiently. This file can be edited as any other text file and it can be used in several programs in The Source's library of business programs to get both current and historical data on securities.

We've also seen that The Source provides a virtual mountain of business information daily giving any business, large or small, an information edge twenty-four hours a day.

CHAPTER 15

Putting It All Together

In bits and pieces throughout this book, we've been seeing some of the wonders of advanced commands. You've learned how to use special text files and command strings to throw your electronic voice across the system. By now, you should feel comfortable with the idea that you can make The Source perform a half-dozen functions with just a few well-chosen keystrokes.

Now it's time for us to keep a promise. We want to show you how to link all of those commands together into a customized system. Before we're finished, you'll know how to:

• Execute entire command strategies with a single keystroke.

• Put The Source on "automatic pilot" so that some operations will be taken care of without your even having to ask.

• Build a personalized menu that reflects your own on-line favorites, a list of your private Top 10 (or 20 or 40 or more) Hot Hits.

• Draw up plans for carrying on the tradition, by further integrating your own hardware and software into a system that takes advantage of The Source's automation.

We're not going to ask you to go on-line in this chapter. However, we bet you won't be able to resist, once you see some of the best tricks in The Source's bag. We hope you succumb to the temptation to log on and play with these new commands after you've finished the chapter. They're fun. More important, they are the key to pushing The Source to its fullest potential and letting you save the most in connect-time dollars.

RECAP: THE FUNDAMENTALS

You already have the basic tools you need.

The first is command stringing. Early on, you learned that there are many ways to get from here to there on the system. For instance, if you wanted to search the POST bulletin board for messages about Apple computers, you *could* walk patiently through the menus. You'd follow the Communication Services branch to the POST program, then select the Scan or Read option, then tell the system you wanted to search by keywords, then say that the keyword by which you wanted to search was "Apple," etc.

But you learned, too, that there's no law against speeding on The Source and you can whip past all the Sunday drivers trudging through the menus. You know how to do the whole business from the -> prompt:

```
************************************************************

-> POST S K APPLE <RETURN>

************************************************************
```

The second tool you've acquired is the command (or CO) file. Discovering it meant we could store entire command sequences in an everyday text file. On a tour, we built a CO file called "GO" that performed four separate operations: (1) turning off the CHAT mode, (2) checking the mailbox, (3) visiting POST and scanning for messages with the keyword of "Apple," and finally, (4) displaying a list of all Apple messages it found that had been posted that day. That whole sophisticated strategy could be executed from the -> prompt by typing:

```
************************************************************

-> CO GO <RETURN>

************************************************************
```

We'll be using CO files here, too, but this time around they'll be pawns, not the brains of the operation.

The third important utility in the toolbox is the Command Output, or COMO, file. As you recall, this is the one that lets you capture material from around the system in a file in your User File Directory. For example, if you type

****

-> COMO NEWSTUFF <RETURN>

****

and then go to read your new messages in PARTICIPATE, all of the material you see along the way will be "photographed" and stored in the file called NEWSTUFF. Then you can return to the -> prompt and type

****

-> COMO –E <RETURN>

****

to turn it off. After that, NEWSTUFF can be read, edited, mailed, or deleted, just as if you had typed it in yourself.

SUPERTOOLS—ABBREVIATED KEYS

Let's hang a few more tools in your workshop.

Not many subscribers use it, unfortunately, but The Source has a powerful option called "Abbreviated Command Set" (or "ABbrevs") that lets you link all of your favorite command strings and CO files to specific keys on your keyboard. For instance, you could define the 1 key to be your "Find all the good stuff on POST" button. Then whenever you wanted to look for the latest on the bulletin board, you could type 1 <RETURN> at the Command Level prompt and away you'd go. In the same set of keys, you might define the 4 as the "Read all the messages in my PARTI INBOX" key, another as the "Send a letter to ol' Dave" key, etc.

This concept probably isn't unfamiliar to you, particularly if you've used some of the sophisticated computers and software available today. Many computer systems let you define "macro keys," which automatically issue several command sequences you've defined. This is the same idea.

Here's how you can set them up on The Source.

First, pick a short name to call the file in which the commands for your keys will be stored—"KEYS" is popular. At the -> prompt, type:

****

-> AB -CR KEYS <RETURN>

****

That tells The Source, "I want to do something with the ABbreviated Command Set (AB). I want to create (-CR) a set called KEYS." The system will mull that over for a second (while it's opening a special KEYS file), then return you to the -> prompt.

From then on, anytime you want to work with the KEYS file—add, change, delete, or list from it—you should start the command with AB. That'll tell the system, "Attention: the following has to do with my ABbrevs."

Let's suppose we'd like the first key to take us to PARTICIPATE to check on new messages. At the -> prompt, we'd type:

****

-> AB -ADD 1 PARTI R <RETURN>

****

Let's dissect that. The first part is familiar—that's the AB command. The last part's not new either—PARTI R is the command string that summons PARTI and reads the messages in our INBOX. The middle part—"-ADD 1"—tells the system what to do with that command string: "wire" it (in a software sort of way) to the 1 key. Note that nearly all the functions that follow this new AB command are preceded by a space and a hyphen.

Maybe the next ABbrev in the KEYS file should let us play the on-line Adventure 550 game. Easy enough:

```
*******************************************************************

    -> AB -ADD 2 PLAY ADV550 <RETURN>

*******************************************************************
```

Then anytime we wanted to play, we could type 2 <RETURN> at the Command Level.

We've been using numerals here, but we don't have to. Perhaps you'd find it easier to remember which key does what if we wired them to letters. The "Play Adventure 550" function could just as easily be assigned to the A key. We could type AB -ADD A PLAY ADV550 <RETURN>. If you wanted the PARTI R button to be the P key . . . you got it: AB -ADD P PARTI R. You can even use ampersands and semicolons if you're into symbolism. And, an ABbrev can be more than one letter—such as PA as your PARTI key. That one would be AB–ADD PA PARTI R <RETURN>. Also, if you use letters, you don't have to worry about upper- and lowercase. If PA is your PARTIKEY, you could execute it with:

```
*******************************************************************

    -> PA <RETURN>

*******************************************************************
```

Or:

```
*******************************************************************

    -> pa <RETURN>

*******************************************************************
```

How many ABbrevs can you have? How many keys are on your keyboard? We have a friend at The Source who has forty keys defined in one ABbrevs set.

And here's an important point: you're not limited to single-line command strings. Instead, you can link complicated CO files into your ABbrevs set as well. We reminded you of our old GO file a minute ago.

If you wanted that routine as part of your KEYS set, perhaps as the 8 key, you'd type

****

-> AB -ADD 8 CO GO <RETURN>

****

meaning that when you pressed 8 <RETURN>, the system would know to execute the entire command sequence outlined in the GO file.

Now, isn't that an interesting development? The number of combinations of creative CO files linked into streamlined ABbrevs is darn near limitless. Hang on—it gets better. First, though, we have to figure out how to make these keys do their stuff.

FIRING UP YOUR KEYS

When you've finished defining your keys (putting in each one with the AB -ADD command), you needn't worry about closing the KEYS file. Your ABbrevs are stored in a specific kind of file, and The Source is taking care of business.

If you want to admire your work, just type:

****

-> AB -LIST <RETURN>

****

The system will display all your ABbrevs, listing the command associated with each one. If you want to delete one of them, say the 3 key, type:

****

-> AB -DELETE 3 <RETURN>

****

If you want to change one: –DELETE it first and then –ADD the new definition. If you want to delete the whole set and start over again, type DEL KEYS <RETURN>, then create a new set with the AB –CR command.

After you get your keys defined the way you want them, all you have to do is "turn them on," by typing:

****

-> AB KEYS –ON <RETURN>

****

And, you guessed it, turning them off again is AB KEYS–OFF <RETURN>.

Once the keys are –ON, they're at your disposal whenever you're at the -> prompt. For instance, if you had defined the keys as we discussed here, to read your PARTI INBOX messages, you'd simply type:

****

-> 1 <RETURN>

****

The system would display the command (PARTI R) and then act on it, whisking you away to PARTI.

In a moment, we'll put together a hypothetical set of ABbrevs and give you some food for thought for your own KEYS construction.

But, first, we have a few more utilities for your toolbox.

PERSONALIZING THE PROMPTS

Ever get tired of the -> prompt and those impersonal error messages you see when you make an on-line mistake? You can change them. The folks at The Source have an option that lets you change your Command Level prompt to whatever you'd like. Sometime when you're on-line, try this:

****

-> RDY –RB 'Howdy!' <RETURN>

****

From then on, your Command Level prompt will be:

****

Howdy!

****

If you wanted to change the error prompt, which is now the rather unimaginative ?> symbol, you could type:

****

-> RDY –EB Darn! <RETURN>

****

Then if you entered a command incorrectly, you'd get a more interesting message:

****

IMPROPER COMMAND NAME (PLEASE REENTER)
Darn!

****

These changes would be in effect for the current session only. If you logged off and back on again you'd be back to the -> and ?> prompts.

And what are these options good for?

Well, uh . . . gee, beats the heck out of us. But then, not everything has to have a profit margin. Just keep them around for those times you're feeling whimsical or want to show off your wizardly ways. And they might be useful if you're helping a neophyte customize his Source account. A prompt that reads "Enter command:" might be more helpful than "->."

BUILDING UP STEAM: THE SID

We've saved the best new tool for last. It's the SID file, and with it you can take control of your own on-line destiny. The SID file will execute the commands you specify each time you log on. It's all automatic—you don't even have to touch the keyboard after you enter your user ID and password.

This is another computer concept you might already be familiar with. Perhaps the operating system you use with your computer lets you "AUTO EXECUTE" or "CONFIGURE" a disk. That means that each time you put that disk in the drive and push the RESET button, a specified program automatically starts running. It's the same idea here, except the SID lets you "autostart" The Source.

At a minimum, the SID file can take care of any routine tasks you always perform right after log-on. Probably the most common use for the SID is to automatically turn off the CHAT mode and check the mailbox, two procedures many of your fellow subscribers run through each time they log on.

You can activate your SID file at the -> prompt by typing:

****

-> ENTER SID <RETURN>

****

Very much like a CO file, the system will respond with an "Enter text" message after which you can type the commands you'd like to have performed automatically at log-on. For instance:

****

CHAT -OFF <RETURN>
MAILCK <RETURN>
<RETURN>

****

Note that, as with most files you create on-line, you close the SID file by tapping RETURN on a new line.

Once you've activated your SID file, the log-on procedure will look a little different. After receiving the usual greeting from The Source and any special bulletins, you will bypass the introductory menu and be taken directly to the -> prompt where the commands outlined in your SID file will be acted upon, one after another.

If you want to remove the SID file, type DEL SID <RETURN> at the -> prompt. If you want to just turn it off sometimes, enter CONTROL P right after you log on. The CONTROL P will override the SID file and you'll end up right at the -> prompt.

EDUCATING SID

The system wizards didn't stop there in designing the SID utility; it's far more versatile than that. In fact, any program on the system can be visited automatically at log-on if you want to specify it in the SID. Many people like to automatically read any waiting SourceMail messages, for instance, so instead of MAILCK, they specify a command string in the SID file: MAIL R UN (for "read previous unread messages").

A word about this, though.

The SID will execute each command you've specified, one line after another, until it comes to either:

1. the end of the SID file, or
2. a natural stopping place on the system, such as the options prompt in SourceMail.

In other words, the SID needs to be back at the Command Level prompt after completing each task you've assigned it. If, instead of the -> prompt, it comes to an option prompt, the SID will stop dead in its tracks and wait for further instructions. This is important in determining the *order* in which you put commands in the SID.

Suppose you want to have two operations performed by your SID file: you want to read any waiting messages in your SMail mailbox and you want to turn off the CHAT mode. If you were to put them in that order in the SID, as

****

```
MAIL R UN
CHAT -OFF
```

****

you'd have a bit of a problem. Experience on the system has taught you that the first command (MAIL R UN) is going to take you to the SMail program to look for new messages. If it finds one, it will display the header and stop on a -More- prompt. And if it doesn't find one, it'll say, "No More Mail" and stop on this option prompt:

****

<S>end, <R>ead, <SC>an, <D>isplay, or <Q>uit?

****

Either way, you're *not* going immediately back to the -> prompt and so the SID file isn't going to reach the "CHAT -OFF" portion of the command list.

There are several ways around this problem.

An obvious one is simply to change the order of the commands. You know from our tours that CHAT -OFF ends up at the -> prompt. After the SID executes that one, it then can move on to the MAIL R UN command with no problem. After completing MAIL R UN, the SID will stop at an option prompt in the SMail program. However, that's no big deal this time, since you've not specified another command after MAIL R UN.

A different solution is to pick another kind of command. We've seen that MAILCK works differently from any other SourceMail commands. MAILCK simply gives you a status of your mailbox and hurries on back to the -> prompt; all the other MAIL strings—MAIL R UN, MAIL SC, etc.—stop at a prompt inside the MAIL program. So MAILCK is quite handy for some applications, such as SID files.

The operative word in both these solutions is "experience." The best way to design sophisticated command files—whether they are CO files, ABbrevs, or the SID—is to walk through them manually, keeping notes of the commands you use and their order. You'll find that any command sequence you can enter manually can be entered in a command file or a set of command files.

A CONSTRUCTION PLAN

All of the software tools—the SID, ABbrevs, CO files, and command strings—can be shaped into an integrated system. When you're ready to design it, you should start your work right where you are now—off-line. There's no use burning up connect-time dollars before you have to.

With a pencil and paper, then, outline what it is you'd like the system to do for you. Decide:

• What jobs will require CO files. After you've designed them, you can log on for a quick on-line test run.

• What keys you want in your ABbrevs set. Write out the command strings as you think they should be. Later you can test them on-line as well, before committing them to a key.

• What functions you want to be implemented automatically each time you log on. Your SID file will be responsible for those duties.

• To get an idea of how to plan a system like this (and, perhaps, to better appreciate just how powerful the SID file and ABbrevs can be), let's build a system together.

CHARLIE AND DAVE'S SUPERKEYS

While we were writing this book, we constructed a dozen keys into a system of ABbrevs and CO files, all hooked together with a simple SID file. To design it, we started out with a wish list: things we wanted the system to take care of while we explored. For instance:

1. Chatting—we wanted to have the CHAT mode turned off automatically when we logged on. However, in case we felt talkative, we also wanted a fast, easy way to turn on CHAT and check to see if any acquaintances were on-line and wanted to talk.

2. Mail and messages—we wanted the system to make it as easy as possible to stay in touch with all of the communications features. That meant having the SID file check the mailbox regularly and display any new messages there. Also, since we wrote a lot of SourceMail to each other during the research, an expressway to send mail to each other would be nice. Finally, we wanted streamlined ways to check on any messages we received through PARTICIPATE as well as to stay on top of things on POST.

3. News—being information junkies, we've always been attracted to The Source's array of news services. But defining all those keywords is time-consuming. Our system needed to put a high priority on automating the searches.

4. Finances—speaking of staying in touch, we wanted a key to let us quickly check the stock portfolio we'd built in the business area.

5. Recreation—and it's not all business. One key would let us play a favorite game, like Adventure 550.

6. File maintenance—there had to be an easy way to turn the keys off.

7. Menu—we wanted to have a nice menu available to illustrate all our keys, something like:

```
*******************************************************************

-:- SpecialKeys -:-

1 CHAT on                     7 Scan POST
2 PARTICIPATE                 8 Play Adventure
3 Check Stocks                9 Mail a letter to Dave
4 Find SFILES                10 Scripps-Howard news
5 UPI news                   11 Washington Post
6 Keys off
                             <H>elp - list keys

*******************************************************************
```

(Put a bookmark on this page. We'll be referring to this menu in a minute. Later we'll show you how to recreate this menu on-line.)

8. Finally, we decided to change that ambiguous -> prompt to something a little more descriptive, like "Command (Enter H for Help)."

SHARING THE LOAD

After we got our druthers listed, we used a pencil and paper to sort out the tasks. We divided it into three columns—COs, ABbrevs, and SID—and listed the jobs that should be performed by each.

Some of the decisions were easy. Obviously the SID file would be best at handling the tasks at the log-on—turning off the CHAT mode and checking the mailbox. We wrote them in the SID column. That's also where we put the job of changing the Command Level prompt to something more logical. That meant we could set all those jobs aside for the time being: the SID file would be the last part of the system we designed.

Looking over the other tasks, three stood out as being perfect candidates for CO files, because they required more than one command string.

One was the "CHAT on" function assigned to key No. 1. We wanted it to turn on the CHAT mode *and* check to see who was on-line. With the pencil and paper, we sketched a CO file like this:

```
****************************************************************

-> ENTER CHAT <RETURN>
Enter text:
CHAT -ON <RETURN>
ONLINE <RETURN>
<RETURN>

****************************************************************
```

Not exactly a model of sophisticated programming, but it would get the job done. Entering CO CHAT at the –> prompt would make it run, and all we'd have to do was assign "CO CHAT" to the No. 1 key.

Another candidate for a CO file was the key that would check the POST bulletin board, and this one would be a little more elaborate. We wanted to search POST's "APPLE" and "INVESTMENTS" categories for any new messages—that is, those POSTed that day. Using what we'd learned in building the original GO file, we wrote this command sequence to be stored in a file called POST:

```
****************************************************************

POST S APPLE INVESTMENTS <RETURN>
N <RETURN>
D <RETURN>
    <RETURN> (that is, one space with a RETURN)
    <RETURN> (again, a space and RETURN)
<RETURN>

****************************************************************
```

Typing CO POST <RETURN> at the Command Level prompt would do the following:

First, it would run the POST program and scan the categories of APPLE and INVESTMENTS (line 1). After finding all the messages in those two categories, the system would prompt for Narrow or Expand the group. Our CO would anticipate that and, in line 2, send an N for Narrow. After that, the bulletin board would want to know how to narrow it (by keyword, category, date, etc.) and, in line 3, our file would send a D for Date.

Then POST would want us to either specify a date or tap RETURN for "today." Line 4 would take care of that by sending a single space on a line, meaning RETURN. Finally, the POST program, having narrowed the group to today's notices, would give us another opportunity to narrow or expand. This time, our file would take the option of tapping RETURN (line 5, another space on a line by itself) to see a list of those still in the group.

That command sequence could be incorporated into our system by linking the command string "CO POST" to key No. 7 on our menu.

Looking over the list, we saw one other operation that would be a good CO file prospect—Option 3, checking the stock portfolio. For that, we built a short CO file called STOX, based on what we saw in the last chapter:

```
************************************************************************

    STOCKCHECK 1 <RETURN>
    (MYSTOX) <RETURN>
    <RETURN>

************************************************************************
```

That, of course, would call up the Stockcheck program and select the first menu option, the stock quotes feature. The file then would anticipate Stockcheck's request for the name of our portfolio file and enter "(STOX)."

Obviously you could link every key in your ABbrevs set to a CO file. To keep things simple here, though, let's assume that all of the other keys we're going to build can be defined in a single command string.

BUILDING THE KEYS

After we designed the CO files, we wrote them on a word processor and saved them on a disk. We planned to save connect time by uploading them to the User File Directory, using some of the techniques described in Chapter 12.

Then we figured out the command strings for the other keys we'd need, still working off-line with our pencil and paper. After that, we were ready for some serious key construction.

We logged on and, at the -> prompt, we uploaded the three CO files we had saved on disk previously. Using the FILETRAN utility, we transferred the files, creating three COs—CHAT, POST, and STOX. Of course, we didn't *have* to upload, we could have typed them in directly. Uploading was just faster.

Following that, we created an ABbrev set called KEYS, by typing:

****

-> AB –CR KEYS <RETURN>

****

Now consult the page you bookmarked—the one that shows our personal proposed menu. It'll make it easier to follow this next part.

As you can see there, the first key we wanted to define was "CHAT -ON," assigned to key No. 1. To wire it to our CO file called CHAT, we typed:

****

-> AB –ADD 1 CO CHAT <RETURN>

****

At the top right-hand column of our menu is the "Scan POST" function, assigned to key No. 7. It's the job of that fancy POST file we just finished. To put it into the KEYS set, we typed:

****

-> AB –ADD 7 CO POST <RETURN>

****

Don't let us get ahead of you, now. You understand what we're doing? We're telling the special abbreviated key set that, when we type

```
****************************************************************

    -> 7 <RETURN>

****************************************************************
```

the system will act *as if* we had entered CO POST and will follow the instructions in that file. Similarly, 1 <RETURN> tells the KEYS set to fetch the CHAT file and act on it.

Note, too, that we don't have to –ADD these keys sequentially—1, 2, 3, 4, etc. In our example, we're –ADDing them from side to side, as they're shown on our menu. But we could enter them in any order we'd like. It makes no difference to the ABbrevs.

We then assigned the third CO file. Our menu showed that "Check Stocks" goes to key No. 3, so we entered:

```
****************************************************************

    -> AB -ADD 3 CO STOX <RETURN>

****************************************************************
```

The next key on the list is "2 PARTICIPATE" and it doesn't have a CO file, so we just entered the command string:

```
****************************************************************

    -> AB -ADD 2 PARTI R <RETURN>

****************************************************************
```

That means that 2 <RETURN> will send us to the PARTI program where the system will display the messages waiting in our INBOX. If we wanted to scan them rather than read them, we could make that "AB –ADD 2 PARTI S."

Across from that key is the one that takes us to the game Adventure 550, so we typed:

```
****************************************************************

    -> AB -ADD 8 PLAY ADV550 <RETURN>

****************************************************************
```

MORE ELABORATE COMMAND STRINGS

Our next command is the one that mails a letter to Dave. The simplest way would have been to put Dave's user ID as part of the definition. We could have made it:

****

-> AB –ADD 9 MAIL S BBP112 <RETURN>

****

However, we wanted to take advantage of that "express" mail command we learned earlier. It's the one that puts a letter at the top of the recipient's mailbox, so we made it:

****

-> AB –ADD 9 MAIL EX BBP112 <RETURN>

****

The next one, key No. 4, would let us search for PUBLIC files on the POST bulletin board. A piece of cake.

****

-> AB –ADD 4 POST S K PUBLIC

****

The next three keys—Numbers 5, 10, and 11 on our wish list—are the news keys we requested. We wanted key No. 10 to take us automatically to the Scripps-Howard News Service (SHNS) and let us define a keyword. By testing it manually, we discovered that when you type SHNS <RETURN> at the -> prompt, you're next shown a menu on which the second item is always "Search by keyword." So we strung it all together as:

```
************************************************************

     -> AB -ADD 10 SHNS 2 <RETURN>

************************************************************
```

Note that your ABbrevs don't have to be a single keystroke. This one, for instance, will be executed with 10 <RETURN> at the Command Level prompt.

On-line experimentation also showed us that with the *Washington Post* database (WPOST), the tenth option on the first menu is the "Search all articles" option, so:

```
************************************************************

     -> AB -ADD 11 WPOST 10 <RETURN>

************************************************************
```

It would access the *Washington Post* and stop to let us define the keyword.

On the other hand, we decided to be quite specific in the next key. We wanted it automatically to search out all stories filed that day on the United Press International business wires that contained the word "computer." You may remember most of this command string from an earlier chapter:

```
************************************************************

     -> AB -ADD 5 UPI N B COMPUTER,,*,SB <RETURN>

************************************************************
```

That would mean that whenever we typed 5 <RETURN>, the system would run the UPI program and select the <N>ational <B>usiness wires from subsequent prompts, define the keyword as "computer," then automatically tap RETURN at the Date prompt to signal "today's news," (that's the ",," part), then choose to Scan Backward from the latest story. (Incidentally, the asterisk is a nifty special shorthand here to mean "the latest story.")

Two more ABbrevs to go. One was a key that would turn the keys off. It's an easy matter—just type AB KEYS –OFF <RETURN> at the –> prompt. But why not let the keys themselves take care of it?

****

–> AB –ADD 6 AB KEYS –OFF <RETURN>

****

Now, what about this last key? How do we get a "<H>elp" screen from The Source? Well, actually, what we wanted was to see a menu of the keys as we've defined them, an on-line menu just like the wish list we've been working from. And it was easier than it seemed. We opened a new text called MENU by typing:

****

–> ENTER MENU <RETURN>

****

Then we typed the list, just like the menu you see on your bookmarked page. (A better idea would have been to design it off-line with a word processor and upload it with FILETRAN or RCV, as we did with the CO files.)

Once the MENU file was created, we defined the <H>elp key by typing:

****

–> AB –ADD H TY MENU <RETURN>

****

Clever. For one thing, this uses a letter instead of a number—H has come to stand for Help throughout The Source. This way, at the –> prompt, H will print out the MENU file.

TEST DRIVING THE KEYS SET

If you really were designing these keys, you'd probably want to stop at this point and give them all a good test drive. If your typing is like ours, you'd probably have made at least one typing error, or forgotten a command, or changed your mind.

To give them a spin, you'd need to turn the set on

****

–> AB KEYS –ON <RETURN>

****

then try each key to see if it takes you where you need to go. You'd notice right away that KEYS was a smart file, that only at the Command Level do the defined keys have special responsibilities. For instance, 9 <RETURN> would take you to mail a letter to Dave. However, once inside the SourceMail program, neither the 9 key nor any other key in the set would have any special properties. Only at the Command Level do the ABbrevs strut their stuff.

If you decided you wanted to change a key, you would get back to the Command Level and delete it with an AB –DELETE command (as in "AB –DELETE 8" to remove the "Play Adventure" key). Then just AB –ADD the revised key.

PUTTING THE CHERRY ON TOP—THE SID

After we were satisfied with the ABbrevs set, we capped it off with a SID file. Looking back over the list of things we wanted the system to do, there were only a few little tasks left:

Turning off CHAT mode.
Changing the –> prompt to the more literal, "Command (Enter H for Help)."
Turning on the KEYS.
Checking the mailbox.

The SID easily can handle all of those duties. All we needed to do was to be careful of the order in which we placed the commands. We commenced the SID file like any other file on The Source, at the –> prompt, typing:

****

-> ENTER SID <RETURN>

****

The system opened the SID file, and after the "Enter text" message, we typed in the commands, one per line:

****

CHAT –OFF <RETURN>
AB KEYS –ON <RETURN>

****

Nothing complicated here. Now we wanted to change the –> prompt, so we typed:

****

RDY –RB 'Command (Enter H for Help)' <RETURN>

****

Note that the message we wanted was enclosed in single quote marks. By the way, there's a 20-character limit on the size of that message.

For convenience, we decided to have the SID automatically display the menu of the keys, so:

****

TY MENU <RETURN>

****

That simply asks the system to type out the text in the MENU file.

Finally, all that was left was to check the mailbox. We'd been careful to make sure all the other commands we'd assigned to the SID up to this point would return us to the Command Level prompt so the next one in line could be executed. We could do that again by using the MAILCK command. However, since we were at the end of the SID, it did no harm if the last command sent it directly into the middle of the SourceMail program. So we used

```
******************************************************************

    MAIL R UN <RETURN>

******************************************************************
```

to have the SID go to the mailbox and read all previously unread messages it finds there.

That made the complete SID a grand five-liner:

```
******************************************************************

    CHAT -OFF
    AB KEYS -ON
    RDY -RB 'Command (Enter H for Help)'
    TY MENU
    MAIL R UN

******************************************************************
```

To close the file, we tapped a second RETURN at the bottom.

So, what does this look like when you log on to The Source? Like you have a chauffeur handling all the predictable decisions for you. As soon as you type in your user ID and password, the system will check to see if you have an SID file in your UFD and, if so, start acting on it. The system will display each command in the SID as it acts on it, something like this:

****

```
Connected to THE SOURCE
> ID BBZ824
Password?
BBZ824 (user 36) logged on Thursday, 29 Aug 85 18:47:08.
Welcome, you are connected to The Source.
Last log-on Thursday, 29 Aug 85 16:08:44.

(C) COPYRIGHT SOURCE TELECOMPUTING CORPORATION
1985.

-> CHAT -OFF

-> AB KEYS -ON
-> RDY -RB 'Command (Enter H for Help)'
Command (Enter H for Help) TY MENU

-:- SpecialKeys -:-

1 CHAT on                  7 Scan POST
2 PARTICIPATE              8 Play Adventure
3 Check Stocks             9 Mail a letter to Dave
4 Find SFILES             10 Scripps-Howard news
5 UPI news                11 Washington Post
6 Keys off                <H>elp—list keys
```

****

```
Command (Enter H for Help) MAIL R UN
No More Mail

<S>end, <R>ead, <SC>an, <D>isplay, or <Q>uit?
```

****

When we type Q for QUIT at the prompt, the system returns to

****

```
Command (Enter H for Help)
```

****

our substitute for ->.

HOUSECLEANING—SOME FINAL DETAILS

Has all of this got you thinking about some interesting possibilities for your own command file systems? We hope so. The opportunites for some sophisticated cooperation between COs, ABbrevs, and SID files are exciting indeed. Experiment with them, and if you come up with a system you think is really interesting, we hope you'll share it with us. We've sprinkled our user IDs all along the way and we list them again in the "Reaching Us On-line" section in the On-line Survival Kit at the back of the book. So write to us and let us know what you've come up with.

And here are some final thoughts to help you in your own constructions.

First, about the system we've made here. For clarity's sake, we used numbers in our ABbrevs, but some subscribers prefer to use letters. Generally, there's no problem at all with that. However, on rare occasions lettered ABbrevs seem to confuse The Source, particularly when the keys are –ON and you enter a command string not on the menu. For instance, we once built a set that used the S key to search the Scripps-Howard News Service (SHNS). However, we found that we then had trouble entering the command string

```
******************************************************************

    POST S K PUBLIC

******************************************************************
```

at Command Level. The system "saw" the S standing alone in that command string and thought we wanted our S ABbrev key. It took us not to POST but to Scripps-Howard. We got around this little problem by adding a "KEYS -OFF" function to the key set. Then we just turned off the keys when we had a new command string to enter. We also could have defined two letters—such as "SH"—for "Scripps-Howard."

Modifying your keys is easy. Once you've created an ABbrevs set (with AB –CR KEYS), you can add keys at any time by entering the AB –ADD command followed by the number or letter of the key you want to define and the command you want to assign to it. To delete, enter AB –DELETE and the number or letter of the key.

You can delete the SID file by entering DEL SID <RETURN> at the –> prompt.

FOOD FOR THOUGHT

You know, with the advent of sophisticated hardware and software, there are some pretty interesting possibilities for integrating your system with the capabilities of The Source.

Imagine—if you had the right equipment (an autodial modem, software capable of automatically logging on, and some kind of clock either in your modem or your computer), you could actually have your computer log on to The Source after you've gone to bed. It could interact with a SID-ABbrevs system you'd designed to pick up the information, save it—in the program's buffer or on a disk, or print it out or put it in a COMO file in your UFD—and then log off again.

Why, there may come a time when there are as many telecomputing automatons running around The Source as there are flesh and blood people. We might even find CHAT requests being answered by, "Sorry . . . I'm not human. I'm just Barry's Buffer. . . ."

A big word of caution, though. Such command systems have to be designed and tested very, very carefully. The shining possibilities of full automation would be dulled considerably if you got up the next morning to find your system had crashed at an option prompt you hadn't anticipated. After about ten minutes of not hearing from your machine, The Source would log you off. (This "time out" is a precaution of the system.) But it would be frustrating to do all that work and not retrieve the information you were seeking.

The secret is, experiment. Walk through your command system manually before you hook it up in ABbrevs and the SID. Remember that many of the programs, like the SourceMail program, will be looking for a Q or QUIT to return to Command Level. By running through the system manually, you can anticipate these commands and include them in your automated system.

And above all, peek over your own shoulder while you're on-line. Notice the features that you visit regularly. Notice when you're entering the same command sequences over and over. These will be your candidates for automation.

WRAPPING UP

In this chapter we learned a number of new system commands, including:

AB –CR, followed by a file name, which creates an Abbreviated Command Set (ABbrevs).

AB ->ADD, followed by a key identifier and a command string, which adds a key to your ABbrevs, such as AB ->ADD 1 MAILCK.

AB –DELETE, followed by a key identifier, which deletes a key from your ABbrevs.

AB –LIST, followed by the ABbrevs' file name, which lists the keys and their assigned command strings.

RDY –RB, followed by a new prompt, which changes the -> prompt of Command Level to a new message, such as RDY –RB 'Yessir?'

RDY –EB, followed by a new prompt, changes the ?–>) error prompt to a new message.

ENTER SID which allows you to create an automatic log-on strategy for your UFD.

CHAPTER 16

One for the Road

We're nearing the end of our tale, but it's only the beginning for you.

You've been introduced to a lot of new material in these pages and maybe even some new concepts of computing. It's time for a commencement address to put them in perspective.

You've learned that The Source is one large interconnected network that lets you retrieve and send information with speed and efficiency never dreamed of a dozen years ago.

You've learned there's nothing to fear. Nothing you do with your computer will break The Source. No one is going to laugh if you make a mistake. There's not even a Big Brother sitting back at headquarters taking down names and IDs of those who stumble.

You've learned that while it's important to know the specific commands to navigate, it's easier to learn the system if you first concentrate on the concepts. For instance, many seemingly unconnected areas of The Source are quite similar, even though the local dialects—the commands—are different. You've found that once you have tucked away in your memory an overview of what's happening, the things going on in the memory of The Source are considerably less confusing, less threatening.

Now you're ready to have fun with The Source, whether your idea of fun is chatting with an on-line friend on Saturday nights or getting instant stock quotations. As computers and on-line networks become more important to our lives, it's only natural for us to begin to take personal computers "more seriously." Good. But that doesn't mean you can't enjoy them, too. After all, decades ago the technology of the automobile became thoroughly integrated into our society, but many of us still find driving satisfying, even relaxing. It's not unusual, then, if you begin to feel the same way about traveling the electronic back roads with your modem.

Perhaps that's our biggest discovery of all: that the key to saving time and money on The Source is to relax. Clock watching doesn't really save time; it just frustrates you. And it's comforting to know that so much of what you need to do with the system can be prepared off-line, before the meter starts running. You can write your letters and notes beforehand, then connect to The Source and quickly transmit them. Because you know the lay of the land, you can even plan your route through the system and its many databases before you ever turn on the machine.

Better still, what you've learned will help you grow with The Source as it continues to evolve. As new features and new commands are introduced, you'll see them, not as isolated mutations, but as a natural extension to the structure with which you're already comfortable.

SOMETHING SPECIAL: CLUBS ON THE SOURCE

Unfortunately, sometimes we don't follow our own advice.

We weren't particularly relaxed when, during the fine-tuning of this book, we learned that The Source was planning to introduce a new feature, on-line clubs. The race was on to include them in our examination.

Meanwhile, the more emotional member of our team, pried off the ceiling by associates and loved ones, was heard to exclaim: "Writing about computer networks is like trying to put out an All Points Bulletin on The Blob. . . ."

We let him wear his lucky hat for a while and go fishing. When he returned calmer, he was ready to continue tackling the challenge of describing a community that's constantly changing, like the star of that 1950s-era monster movie classic.

And we're glad we did take a closer look. From what we've seen of the prototypes, the new on-line clubs will (1) herald a new era for quality communications features on The Source, and (2) be quite easy for you to learn, because you're already familiar with most of the components.

The Source's reasoning behind these groups is that birds of a feather like to download together. People with similar interests, such as owning the same kind of computers, like to have a corner of the system to call their own. They can kick their shoes off and get the latest news and rumors about their common interests, communicate with each other privately or as a group, and generally feel as if they are part of an on-line club.

So, as you prowl the system in the future, you're likely to see a menu item for these groups. Some of them will be "computer specific," perhaps, dealing with hardware and software subjects. Others will be more general in subject matter.

If you see a menu option for these clubs, feel free to explore—you're well prepared. That's because Source officials tell us they intend to create the new clubs from services and utilities you've met elsewhere in your travels.

For example, a club's menu selection labeled MEMBERS will likely let you search for names and interests of other regulars in the group—just like the Member Directory we met back in Chapter 5.

Newsy notes for and by members are likely to be published and searched in the style of the POST bulletin board program.

The same holds true with the conferencing selection of the groups. You can expect to see something like the PARTICIPATE program available for your electronic conversations with other members.

So, where's the difficulty in navigating an interest group? With the knowledge you've now acquired, there isn't any.

When you think about it, this idea of integrating other program resources into an on-line group is praiseworthy. It makes maximum use of a very fine system. It keeps veteran subscribers learning new tricks. Best of all, it helps newcomers learn other areas of The Source by "taming" seemingly complex systems in a sort of electronic microcosm.

So, what we thought was going to be a logistics problem for us turned out to be a compendium of things we had already described—and a fine example of what the future holds. Because you know the major concepts of The Source you already understand its newest feature.

A FEW OF OUR FAVORITE THINGS

One of the satisfying things about being a member of The Source community is sharing the on-line treasures you've discovered. Nightly, the POST bulletin board, SourceMail, and other communications features carry personal news flashes such as, "Hey, Henry—did you get the word? There's a new conference on foreign cars going on over on PARTICIPATE. . . ."

We also have some gold to share: a few of our own favorite things on The Source. We think you'll find some of them as fascinating as we have. In fact, some of these features slowed us down while we were writing. We'd get involved in the material offered by these areas and forget why we were there.

THE BEST OF PARTICIPATE

We told you about a few of the conferences you'll find valuable when it comes to learning about PARTI. We didn't, however, mention some of the conferences that have attracted our attention for other reasons. Of course, there's no guarantee that these conferences still are going on as you read this—as you know, conferences that outlive their usefulness are often deleted—but we suspect most will live on for a long time.

"BLINK ME"

Steve Gibson is one of the regulars on PARTI and his conferences always are heavily read and answered. In this conference, Gibson has come up with an express method for uploading to the PARTI system any material that has been written off-line. Proponents say it can cut by as much as 90 percent the time it takes to post answers and messages.

Remember how we told you that you can issue nearly any of PARTI's commands at the ACTION prompt? That includes sending messages and notes to other conferences and users.

What if you were able to anticipate all the commands necessary to upload a bunch of messages, one after another? You could assemble all the text files and commands in one large file off-line, then connect to the system and send it to PARTI in a burst.

Gibson calls that method "blinking" because, he says, he does all his uploading in the blink of an eye. The process not only involves PARTI and your own file transfer program, but also some special command for Telenet to help in the uploading.

Before devising his "blinking" system, Gibson said he was spending nearly $900 a month in connect time, most of it in PARTI. With "blinking" he reduced his monthly bill to about $100.

If you're becoming a PARTI animal, it would be worth your while to download the easy-to-understand "BLINK ME" tutorial.

"SPACE"

It's called an "electure"—short for "electronic lecture"—and this conference by Dr. Paul Levinson is one of the most thought-provoking pieces of on-line writing you'll find. Levinson has brought together the human aspect of space exploration: what it already has meant for our species and what it is likely to mean in the future.

"Just as a baby must leave the womb to be satisfyingly human," the conference begins, "just as most folks leave the neighborhoods of their birth as they move through life, so the human species must leave the womb of the Earth to achieve its full human destiny."

What makes this electure different from a classroom lecture is that the listeners have a chance to join in the discussion. To date, subconferences of "SPACE"—including comments from writer Stewart Brand, former astronaut Rusty Schweickart, and science-fiction writer Sylvia Engdahl—have dealt with the possibilities of intelligent life in outer space, traveling faster than the speed of light, even a proposal for a "space tax" to be used solely for space exploration, and more.

"SUPPORT GROUPS"

Groups of people that have formed for mutual support when the going gets tough have become extremely popular in the United States. The electronic equivalents of these support groups are alive and well on PARTI under the general conference called "SUPPORT GROUPS."

When we last checked, the subconferences listed under this main conference included "CONSUMER PROTECTION," "SPLITTING UP," "HOT LINE," "WEIGHT WATCH," "TOUGHLOVE," "STOP SMOKING," "SEXUALITY," and "SINGLE PARENTING."

The most active, by far, was "WEIGHT WATCH" where people who are trying to lose weight gather to confer about their dieting problems.

"RELATING—FOR YOU"

This conference, and its related subconferences, is a place for relaxing conversation. At times it is very personal. Sometimes, it's just good-natured, friendly conversation that can make you feel almost like part of a family.

NEWSBYTES

Moving to other areas of the system, San Francisco journalist Wendy Woods has become an on-line success story on The Source. Her "Newsbytes" newsletter started out a few years ago as a self-published journal of computer-industry news. It developed along the usual lines, from Sharefiles to Member Publishing.

Now Newsbytes has been so successful that The Source has moved it to its own special corner of the system. Woods and her friends and correspondents in England, Canada, and Japan keep you posted on the world computer scene with a regular weekly newsletter, as well as news flashes throughout the week. For instance, the Newsbytes crew often is on the scene of a major computer show to post around-the-clock bulletins in a special area of the publication.

To reach it, type NEWSBYTES <RETURN> at the –> prompt.

"GOODIES"

"BLINK ME" star Steve Gibson is also a computer hardware developer and has produced an excellent on-line newsletter which contains all sorts of industry tidbits. They range from outright gossip about the computer industry to hard facts about the newest models of computers, software, and peripherals. Gibson has scored more than one "scoop" in this conference since it began in February 1985. Many of the things you'll see in this read-only conference may be published in the computer magazines as news a few weeks later.

A companion to this conference is "GOODIES TALK," where you can make comments about Gibson's reports and ask questions of him and other "GOODIES TALK" joiners.

MICRO MONITOR

Columnist John Dvorak has gained a reputation for being up front and direct with his thoughts and comments on the computer industry. Thousands have come to know him through his column in INFOWORLD. As we were in the final stages of writing this book, it was announced that his weekly "Micro Monitor" column is to be featured on The Source. Dvorak told us he wanted his column to be available at the Command Level prompt by typing MONITOR.

HIGHLIGHTS OF MEMBER PUBLISHING

Speaking of Member Publishing, there are some excellent journals and newsletters on-line for you. As we've said, to keep up-to-date on the latest, you should visit the Publication Index option from time to time.

And here are some of our personal favorites:

The New Tech Times

This is a newsletter published by the people who produce *The New Tech Times* public television program. It's meant as a supplement to the broadcasts and gives information about upcoming shows, air times, background information on guests, etc. To reach it, type PUBLIC 125 DIRECT at the -> prompt.

Real-Times Magazine

For those who follow the computer industry, this is a stand-out publication with regular commentaries by industry observers like writer Larry Madgid. It usually focuses on what's new and hot in the world of hardware and software. Reach it by typing PUBLIC 111 DIRECT.

Comedy by Wire

Sometimes our critics say the computer world is humorless. Don't tell Billiam Coronel that. He publishes a weekly comedy magazine which he's dubbed, "The second choice of a new generation." Coronel became an "underground" hit when he was publishing his work in his Sharefiles. His new Member Publication has made him the subject of reports in the *Wall Street Journal*, United Press International, and other national publications. To look in on it, type PUBLIC 153 DIRECT.

Apple City

Walt Marcinko's Apple City has served for more than three years as a pillar on Member Publishing row, a wellspring of current information about Apple computers, programs, and related products. It's also a good example of how some publishers customize their products with their own menus and options. To visit, type PUBLIC 113 DIRECT. If it's not there, it may be this service has become one of the new clubs we mentioned earlier in this chapter.

Public Access Systems

If you're an enthusiast of computer bulletin-board systems (BBSes), you don't want to miss this publication, provided by Robert Neibiker of Athens, Ohio. It's a searchable database of BBS numbers across the country, based on the Public Access Message Systems (P.A.M.S.) list compiled by Bill Blue of Santee, California. It will allow you to search for BBS numbers anywhere in the country and to add your own. To reach it, enter PUBLIC 112 DIRECT.

CompuDate

This publication, provided by a group called H.U.M.A.N.S., Inc., is living proof that computers talk about more than computers. This is a "nationwide computerized single's club" which specializes in on-line matchmaking. It's a members-only publication—you'll be asked to pay a membership fee. That would entitle you to file your own profile on-line and search the database for people of like interests. To check it out, type PUBLIC 122 DIRECT. The publication will give you general instructions, even if you're not a member.

THIS IS NOT GOOD-BYE

Before we started our journey together, you may have perceived this computer system with the authoritative-sounding name as some sort of giant electronic superintelligence machine from which all knowledge flowed on one-way channels.

Happily, we've found that's not the case. Because of its interactive nature, The Source is as much *you* as it is anyone, or anything, else. The system thrives, not on its own imagination, but yours. The fact that The Source continues to evolve is evidence that the communications is working.

In other words, we've come all this way to find that The Source, for all its state-of-the-art communications facilities, is a network of people.

What does that demand from you? Nothing new, really. If this is just the latest edition of an age-old social contract, then you already know the rules of being a good citizen:

Try to enhance this exciting medium by sharing what you know with someone else. (That'll make you feel a little better about it the next time you are the one asking for directions.)

Be tolerant of others, whether they are newcomers or veterans, and more than likely they'll be tolerant of you.

Report abuses of the network. It doesn't mean you must put on your policeman's hat and make an electronic citizen's arrest. Just apply the same good standards you have for your own neighborhood.

Above all, contribute, and you'll find the art of conversation isn't dead at all.

Enough philosophizing. It's time you moved out on your own. But we won't say good-bye. We're travelers of the same system you're exploring and as close as the SourceMail mailbox. Our user IDs are listed in the On-line Survival Kit that begins on the next page.

Enjoy yourself—and drop us a line if you're in the neighborhood.

The On-line Survival Kit

Now that you're heading out on your own to explore The Source, we want to leave you with a going-away present, a kind of electronic first-aid kit that you can tuck under the dashboard of your computer for emergencies.

It's a grab bag of commands and features of the service that you can have at arm's reach for quick reference. We've divided it into seven parts:

I. Hooking Up—how to find phone numbers and set communications parameters.
II. Command Compendium.
III. Getting Help and Helping Yourself.
IV. The Address Book—where to find some of The Source's major features.
V. The Business of Business—helpful information in using The Source's financial and business features.
VI. The Bookshelf—further reading.
VII. Reaching Us On-line.

I. HOOKING UP

The Source has a database of current access numbers throughout the country—particularly useful if you're hitting the road with your trusty portable, to be checking in from parts unknown.

To use it, type ACCESS at the Command Level -> prompt and the system will display:

****

LOOK UP ACCESS NUMBERS BY:

1. CITY (OR CITY & STATE)
2. STATE
3. AREA CODE
4. SEE ALL ACCESS NUMBERS
5. SIGN-ON INSTRUCTIONS

<I>NSTRUCTIONS

ENTER ITEM NUMBER OR <H>ELP:

****

Communications Parameters

The Source recommends you configure your communications software this way:

1 stop bit
8 data bits
No parity
Full duplex
300 or 1200 baud

II. COMMAND COMPENDIUM

As you prepare for your independent study of The Source, we'll leave you with a quick reference for some of the major commands of the system. This section places particular emphasis on CONTROL key commands, Command Level commands (those you can enter at the -> prompt) and some additional commands for the powerful SourceMail service.

CONTROL Codes

CONTROL keys are very useful for freezing and unfreezing displays so that your eyes can keep up with your modem. And they are the ticket to sending The Source the message, "Naw, I don't want to read this—stop and take me back to the Command Level prompt."

Here are the CONTROL codes used on The Source:

CONTROL S—freezes the display. This is handy if the information is coming in too fast to read and you want to say, "Hold up!"

CONTROL Q—unfreezes the display after you've entered a CONTROL S.

CONTROL P—can be a BREAK on The Source. It interrupts the program and takes you back to the primary prompt or to the Command Level.

CONTROL H—is the backspace, just in case your communications program doesn't have backspace built in.

CONTROL G—sounds a bell on your computer if it has sound capabilities. Not all systems allow this one.

CONTROL L—clears the screen on some terminals, or can cause a form feed on a printer.

CONTROL E—in The Source's CHAT mode, displays the account number of the person to whom you're chatting.

Command Level Commands

Menus are fine, but Command Level commands are quicker. Here are some of the major commands used on The Source. These can be entered at the Command Level -> prompt:

AB -CR—creates a file of abbreviated commands.

ALC—moves you up to your log-on User File Directory from any subdirectory.

ASF—attaches you to your Sharefiles area.

CHAT i.d.—allows real-time conversation with another person on-line.

CHAT -OFF—automatically turns down any CHAT requests.

CHAT -ON—accepts any CHAT calls.

CO file name—executes a command file.

COMO file name—preserves screen display in a file.

COMO -E—turns off COMO, that is, ends a COMO session.

CPW old password—changes your password.

CRTLST file name—displays a file, pausing every 24 lines.

DATE—displays time and date.

DEL file name—deletes a file.

ED or ED file name—uses the text editor.

ENTER file name—opens a file.

F—a file directory, displays your file names and amount of record storage they occupy, with the date they were last modified.

FILETRAN—uploads a text file from your micro to your User File Directory (UFD).

GETSHR—copies a shared file to your Sharefiles area.

HELP program name—gets information on how the program operates.

ID—followed by your account number and password, signs you on to The Source.

L—lists the titles of files in your User File Directory.

OFF—logs you off the system.

ONLINE—lists the IDs of members currently on-line.

PUTSHR—moves a file into your Sharefiles area.

RCV—uploads a text file from your micro to your UFD.

RDY -RB new prompt—changes the Command Level symbol (->) on your account.

TIME—displays time used during current session.

TY file name—displays a specified text file.

TYPSHR—displays a shared file belonging to another member.

WHO—indicates ownership of account and location among directories and subdirectories.

Building SourceMail Mailing Lists

During our tours, you learned how to use SourceMail for day-to-day correspondence, and even how to store a form letter in your User File Directory for use in SourceMail with the dot-LOAD command. The SMail program also can accommodate bulk mailing of electronic mail.

If you wanted to send an SMail to more than one person on the system, you could laboriously type in all the user IDs at the "To:" prompt, one after another, separated by spaces, such as BBZ824 BBP112 TTT111, etc.

However, The Source has two handy shortcuts.

One is a simple mailing list in which you store the user IDs of all the people in your group. Let's pretend you're creating a mailing-list file called "FRIENDS." You could open the file at the -> prompt

****

-> ENTER FRIENDS <RETURN>

****

and then type in each user ID, one at the beginning of each new line, such as:

****

BZZ000 <RETURN>
BZZ001 <RETURN>
BZZ002 <RETURN>
<RETURN>

****

To use your new mailing list, all you'd have to do is go to the MAIL program and, at the To: prompt, type

****

To: ML FRIENDS <RETURN>

****

telling the system, "use Mailing List FRIENDS from my UFD." However, an even quicker route is to string together the commands at the Command Level prompt:

****

-> MAIL ML FRIENDS <RETURN>

****

Once you write the letter and send it (with dot-S), it automatically will be mailed to everyone on the list.

Meanwhile, old pros around The Source use a more flexible bulk-mailing system: the distribution list. It's similar to the mailing list, except that it can keep more than one list in the same file, with up to two hundred user IDs on each list. Think of it as your electronic little black book of addresses.

The Source requires a specific file name for the distribution list—MAIL.REF—so to start one, you need to create a file by that name. At Command Level, you would type:

****

-> ENTER MAIL.REF <RETURN>

****

Then you would type in your lists flush left (no indent) on each line, beginning each list with a title for that group. For instance:

****

FRIENDS BZZ000 BZZ001 BZZ002 BYZ123 BZY321 <RETURN>
CHARLIE BBZ824 <RETURN>
DAVE BBP112 <RETURN>
MUSIC TTT111 TTT222 TTT333 <RETURN>
<RETURN>

****

Some lists might just contain a single name, as in the CHARLIE and DAVE lists. Others can have up to two hundred names in them.

A word about that, though: if a list is longer than 80 characters, just tap RETURN at the end of the first line and begin the next line with the & character (ampersand), such as:

****

MUSIC TTT111 TTT222 TTT333 TUU111 TTU 121
& TZZ000 TYZ101 TRR333 TRT444 TWW000
& TPP101

****

Here the MUSIC list is more than 80 characters long, but the ampersands are used to break it up into three lines.

You also can include in the list some special commands to the SourceMail program—QUIET and NOSHOW—as in this list:

****

CLUB NOSHOW SSS101 RSR666 YYY111 WWW777
& JJJ555 KOO222 QUIET

****

The NOSHOW at the beginning of the CLUB list will tell the SourceMail program *not* to show each recipient the user IDs of everyone else who received the letter. This makes the message seem a little more personal for the recipient. More than that, it's a courtesy that saves the user a little connect time—without the NOSHOW, each recipient would see a long list of user IDs before getting into the letter itself.

The QUIET on the other end of the list saves you some time. It tells SourceMail not to bother listing all the user IDs for you after you've typed .S <RETURN> when mailing the letter.

As always, you close the MAIL.REF file with a second tap of the RETURN key.

Sending bulk mail with this super distribution list is similar to using the single mailing list we've just seen, except that instead of signaling with ML, you put the list's name in parentheses. After going to the MAIL program's Send option, you could address a letter to your FRIENDS group with this:

****

To: (FRIENDS)

****

Or, if you wanted to mail directly from the Command Level:

****

-> MAIL (FRIENDS)

****

Just remember that the parentheses will tell SourceMail, "Look for a list by this name in my MAIL.REF file," while ML followed by a file name says to look for a single mailing-list file by that name.

III. GETTING HELP AND HELPING YOURSELF

The Source has help files throughout the system. As you've seen, entering the word HELP at many of the system's prompts will give you additional information on the options available to you at that point.

Some Useful Addresses

In early 1985 The Source introduced a new, free user help program called INFO. (You met it on your first tour of The Source back in Chapter 3.) Questions and answers in this area of the system are regularly updated so, since this part of the service is free, it behooves you to check in from time to time for the latest from the management. Just enter INFO at the –> prompt for this menu:

```
*****************************************************************

   MEMBER INFORMATION <INFO> **FREE**

    1 Using The Source <INTRO>
    2 Contacting The Source
    3 How to Change Your Password
    4 Rates and Storage Fees <RATES>
    5 Billing Questions
    6 Frequently Asked Questions
    7 Membership Agreement
    8 Ordering Additional Accounts
    9 Ordering The Source Manual
   10 How to Become a Member Publisher

*****************************************************************
```

Some of the options on this menu are of particular interest because they give you direct (free) access to various departments in The Source. For instance, to order additional accounts, select Option 8; to order The Source manual, it's Option 9; for the billing department, you would enter INFO and select Option 5 from the menu.

Speaking of billing, if you sometimes need to change information pertaining to your account, you can write SourceMail to TCA063.

And, if for any reason you want to cancel an account, you can write on-line to TCA064. Or off-line to The Source, Cancellations, Billing Department, 1616 Anderson Road, McLean, VA 22102. (Allow about thirty days for the cancellation to take effect.)

There are other important on-line addresses you should know about.

Entering SUPPORT at the Command Level prompt will take you to the customer-support department's special mailbox. If you need to contact The Source customer support by phone, call 800-336-3330 or 703-821-8888. The staff's on duty daily from 8:00 A.M. to 1:00 A.M. Eastern Time.

Have a suggestion for changes in the system? Enter SUGGEST at the prompt.

Here are some other frequently used Source mailboxes:

MANUAL: to order the user's manual
SOURCE: Investor Services mailbox
STCPUB: Member Publishing
STC320: information about private networks
TCA068: fulfillment (manual, SourceLink orders copy of the system's menu map.)
TCA088: customer support
TCA098: public relations

You and Your Password

Your password on The Source is like your house key and your checkbook. Protect it. Don't tell it to anyone on the system. Don't leave it beside your computer. And change it regularly.

A reminder on how to do that. At the Command Level -> prompt, enter CPW <RETURN>. You'll be prompted for the old (that is, current) password, then for the new password. The Source will ask you to confirm the change by typing the new password in a second time.

If something should happen to your password—if you forget it, or if you have reason to believe it's been stolen—contact The Source's customer-support people immediately. They will want to know your name, address, telephone number, credit card number, and your mother's maiden name. (What? Yes, if you'll recall, that was one of the questions on your application for a subscription. This was why—to verify on the phone that you are who you say you are.)

After verifying this information, you will be issued a new password.

The Bill

The Source's rates are subject to change, so it's not a bad idea to periodically check into the RATES files. The Source is going to notify you if there's any major change in the cost of using the system. Nonetheless, if you don't know whether a feature is surcharged or whether a certain hour is prime or non-prime time, enter RATES at the Command Level -> prompt and read the latest from the management.

Connect-time Charges

In the summer of 1985, The Source introduced a new rate structure and also announced that 2400-baud communication would be available. Here are the connect charges per minute:

	300 baud	1200 baud	2400 baud
Non-Prime Time	14 cents	18 cents	20 cents
Prime Time	36 cents	43 cents	46 cents

Prime time is 7:00 A.M. to 6:00 P.M. Monday through Friday (that's your local time). Non-prime time is all other hours and holidays (usually New Year's, Labor Day, the Fourth of July, Thanksgiving, and Christmas).

The Source charges a minimum of $10 a month.

Storage Charges

The Source also charges you for the amount of information you've stored in your User File Directory and Sharefiles areas.

Here's the latest schedule (mid-1985), based on the number of "records" stored (where one record equals about 2,048 bytes or characters):

1–10 records	50 cents/record/month
11–99	20 cents/record/month
100–999	15 cents/record/month
1,000–9,999	10 cents/record/month
10,000–99,999	5 cents/record/month

Incidentally, storage rates for MAIL.UFD files are charged differently. For information, contact customer support.

Tip: Storage charges are calculated at midnight Eastern Time each evening, so if you do not wish to be charged for storage, delete your files before midnight Eastern Time. Once you have deleted your storage, log off then back on again to initialize the deletion prior to midnight.

Surcharged Services

As you've seen, some features on the system carry extra charges.

As we wrote this book, connect-time rates for these "value-added services" were $39.75 an hour during prime time and $34.75 in non-prime time. And the surcharges for 1200 baud service are additional.

All the surcharged services are noted on the menus. As we wrote this, they included:

"Stock Analysis & Tables" in the Media General service (STOCKVUE).
Airline Schedules and Fares from the Official Airline Guide (OAG).

IV. THE ADDRESS BOOK

Throughout this book we've relied on The Source's excellent menu structure to provide an overview of the system and what it has to offer. But we've also emphasized that running the system from the Command Level prompt is the best way to save time and money.

Some Command Level Addresses

For future reference, here's a summary of major Command Level addresses, organized into interest groups.

Arts, Leisure, and Recreation

Best-seller List	FOCUS
Book Ordering Service	BOOKS
Games Library	HELP GAMES
Movie Reviews	MOVIES and FOCUS
Music to Order	MUSICSOURCE
Radio Programs to Order	RADIOSOURCE
Ski Reports (winter months only)	HELP SKI
Sports Schedules, Scores, and Top Stories	SPORTS
TV Listings	FOCUS

Business and Finance

Finance: Latest Stock and Option Quotes	INVEST and STOCKCHECK
Stock Market Reports	HELP UNISTOX
Stock Performance Analysis	STOCKVUE
Computer Stock Quotes	PCSTOCKS
New York Stock Exchange Indexes	NYSE
Standard & Poor's Indexes	SP
Dow Jones Indexes	DOW
Metal Market Activity	METALS
Gold Prices (Daily)	GOLD
Financial Modeling/Business Planning	HELP MODEL1
Business News (updated throughout the day)	BIZDATE
Business Publications Abstract Service (Management Contents, Ltd.)	CONTENTS
Business Services Index	BIZDEX
Commodity News	CWN, BIZDATE
Donoghue Moneyletter	DONOGHUE, BIZDATE
Employment Network	EMPLOY

Communicating with Others

Bulletin Boards	POST
CHAT (live conversation with another member of The Source)	HELP CHAT
Classified Ads	POST
Conferencing	PARTI
E-COM Mail Message Service	ECOM
Mail	MAIL
Mailgram Message Service	MGRAM
Members Currently On-line	ONLINE
Member Directory	MEMBERS
Teleconferencing	PARTI
User Publishing	PUBLIC

Computer Information

Digital Research, Inc., Information	DRI
Microcomputer Information	MICROSEARCH

Daily News and Research

Almanac (daily)	FOCUS
Associated Press Select News Stories	AP
Editorials (syndicated columnists)	BYLINES, WPOST, and FOCUS
Federal and Government News	WPOST
Horoscopes	FOCUS and BYLINES
News Bulletins and Top Stories	BULLETIN
News Features	FOCUS and BYLINES
Personality Spotlight	FOCUS
Research Service (Information on Demand)	HELP IOD
Scripps-Howard News Service	SHNS
System News	NEW
United Press International News Wire	UPI
Washington Post Select Stories	WPOST
Weather Forecasts	WEATHER

Shopping

General Merchandise	CSTORE
Records and Tapes	MUSICSOURCE
Classic Radio Programs	RADIOSOURCE
Book Ordering	BOOKS

Travel

Travel Services: Airline Schedules	HELP OAG and HELP AIRSCHED
Hotel Guide (worldwide)	AZ
Hotel Guide (U.S. and Canada)	HELP USROOM
Restaurant Guide	HELP USREST
Travel Club Info & Travel-related Services	FIRSTWORLD
Vacation Packages	TOURS
Register with Travel Club	CINFO
Make Reservations	LETSGO
Weather Reports	WEATHER

Various Things about the Source

Announcements on The Source	NEW
Commands Used on The Source	HELP COMMANDS
Encode Text Files	HELP CIPHER
Menu of Services on The Source	MENU
Phone Numbers to Connect with The Source	ACCESS
Programming/Advanced Applications	HELP ADAPPR
Rates and Fees for The Source	RATES
Shared-File System Explanation	HELP SHAREFILES
Text Editor	HELP ED
Using The Source	INTRO

Tips on Reading Member Publishing

All of The Source's user-published newsletters are introduced by five or six lines of disclaimer. If you'd like to avoid that disclaimer on your fast-lane commands, just add the word DIRECT to the end of your Command Level -> command. For instance, such as PUBLIC 123 <RETURN>.

In the world of electronic publishing, as in publishing everywhere in every medium, publications come and go. If you're a user-publishing enthusiast, you should regularly collect the latest index of publications. Just enter PUBLIC 7 at the Command Level prompt.

Here's the latest list of publications available as we completed this book:

Title	Location Number	Publisher
AINetwork	119	(BBJ081)
Anne's Annals	159	(AAB470)
Apple City	113	(TCD912)
Christology Course	124	(STI016)
Comedy by Wire	153	(BBJ906)
Dial-a-Date	122	(TCA550)
Digital Decisions	118	(BBF611)
Eater's Digest	156	(STY542)
Elephant Walk Computers	172	(BBT394)
First Byte Software	117	(STX313)
Gourmet Diner's Club	127	(STQ967)
Image Micro Catalog	171	(ST7367)
Independent Record Releases	175	(STX453)

Jobhunter's Handbook	126	(STV709)
Kaypro-Knews	182	(BBL126)
Micro City	176	(TCP092)
The Muses	151,114	(TCP831)
NEC Advanced PC Info	183	(BCE891)
New Tech Times	125	(BBI599)
Online Educator	157	(ST6588)
Oxbridge Legal Forms	123	(STN250)
Product Reviews	116	(TCY617)
Public Access Systems	112	(TCU583)
Real Times Magazine	111	(TCS091)
Religion Online	128	(BCF060)
S.A.U.G. Magazine	115,133	(TCA265)
S.A.U.G. Libraries	22	(TCA265)
S.I.G.O.P. Library	21	(TCV176)
Subfile99	181	(TI5361)
Teleresource	174	(ST3899)
W.I.N.K magazine	154, 131	(TCV176)

V. THE BUSINESS OF BUSINESS

The Source has many fine business and financial services. Some require special codes and addresses all their own.

Management Contents

Management Contents, Ltd. (CONTENTS) provides abstracts of hundreds of business journals and magazines. To search specific publications, you must use a three-letter code. Here's the list as of mid-1985, sorted by the publication name instead of the code.

Code	Publication Name
ACR	Accounting Review
CBR	Across the Board
ADV	Advertising World
AMD	American Demographics
AER	American Economic Review
AEC	American Economist
ASM	Association Management
BKR	The Banker
BAR	Barron's

BIH	Best's Review Life/Health Insurance Edition
BEN	Black Enterprise
BUS	Business and Society Review
BEC	Business Economics
BHO	Business Horizons
BWE	Business Week
CBM	Canadian Business
CJE	Canadian Journal of Economics
CWM	Cashflow Magazine
CHS	Chain Store Age Executive
CFM	Credit & Financial Management
DIM	Direct Marketing
DBM	Dun's Business Month
EUS	Economic Outlook USA
EKC	Economic Review—Federal Reserve Bank of Kansas City
ECN	Economist
EBU	Electronic Business
FHL	Federal Home Loan Bank Board Journal
FNY	Federal Reserve Bank of N.Y. Quarterly Review
FSL	Federal Reserve Bank of St. Louis Review
FED	Federal Reserve Bulletin
FIA	Financial Analysts Journal
FEX	Financial Executive
FPL	Financial Planning
FBR	Forbes
FOR	Fortune
FUT	Futurist
GAO	GAO Review
HBR	Harvard Business Review
HFM	Healthcare Financial Management
IDE	Idea: The Journal of Law & Technology
INC	INC
IDW	Industry Week
IRM	Information Management
ISU	Information Services and Use
IIN	Institutional Investor
III	Institutional Investor International Edition
IEC	International Economic Review
JAA	Journal of Accounting, Auditing & Finance
JOA	Journal of Advertising
JAD	Journal of Advertising Research
JCE	Journal of Comparative Economics
JCA	Journal of Consumer Affairs
JFI	Journal of Finance

JFM	Journal of Futures Markets
JMK	Journal of Marketing
JRB	Journal of Retail Banking
JRL	Journal of Retailing
LRP	Long Range Planning
MMQ	Management Quarterly
MED	Marketing and Media Decisions
MCM	Marketing Communications
MRW	Marketing News
MES	Medical Economics
MER	Merchandising
MAC	Mergers & Acquisitions
MHC	Modern Health Care
MON	Money
UML	Monthly Labor Review
NAB	Nation's Business
OFF	Office
ADM	Office Administration and Automation
PIA	Pensions & Investment Age
PTA	The Practical Accountant
PRJ	Public Relations Journal
RAA	Real Estate Appraiser & Analyst
SAL	Sales & Marketing Management
SMR	Sloan Management Review
SBR	Small Business Report
SCB	Survey of Current Business
TTA	The Tax Advisor
TAM	Taxes, The Tax Magazine
TCR	Technology Review
VEN	Venture
VIT	Vital Speeches of the Day
WMB	Women in Business
WWM	Working Woman

Career Network

In The Source's Career Network (EMPLOY) feature, the database asks for a "job category code." You can have the system list those forty categories. Here was the list as of mid-1985:

ACC Accounting
ADM Admin/Personnel
ADV Advertising
AER Aeronautical
AGR Agricultural
ART Arts

BANK Banking/Invest
COMM Communications
COMP Computers
CON Construction
CONS Consulting
ECO Economics
EDU Education
ELE Electronics
ENE Energy
ENG Engineering
ENT Entertainment
FIN Finance
HOT Hotel
IND Industrial
INS Insurance
LAW Law
MANA Management
MANU Manufacturing
MAR Marketing
MAT Mathematics
MED Medical/Pharm
PUB Publishing
REA Real Estate
RET Retail/Wholesale
SAL Sales
SCI Sciences
SER Service
SPO Sports
TEC Technical
TEL Telecommunic.
TRAN Transportation
TRAV Travel
UTI Utilities
MIS Miscellaneous
GRAD Recent Grads

Reading STOCKVUE: The Screens

The stock analyses provided by the Media General Financial Services database called STOCKVUE is subdivided into groupings called "screens." The following is a list of the screens and their constituents:

Screen 1—"LOCATOR"—shows stock name, symbol, industry group, exchange, and a description of the industry group.

Screen 2—"INDUSTRY GROUPS"—for any given group number, shows stock name, symbol, exchange, current price, and percent of change this week for each stock.

Screen 3—"PRICE CHANGE & RANGE"—shows weekly high, low, closing value and percent of change, 200-day moving average, and 52-week high and low.

Screen 4—"PERIODIC PRICE CHANGE (%)"—shows percent of change in last trade week, last 4, 13, and 52 weeks, and year to date.

Screen 5—"PRICE EARNINGS RATIOS"—shows current and 5-year average high and low price-earnings ratios, and 5-year high, low, and current price.

Screen 6—"RELATIVE PRICE/ACTION"—shows market value as a percent of revenue, percent of market and industry norms, price equity ratio, relative price index, earnings yield, up beta, and down beta.

Screen 7—"TRADING VOLUME"—shows shares traded, volume as

a percent of shares outstanding, weekly dollar value of shares, liquid. Ratio, and on-balance index.

Screen 8—"TREND TO MARKET"—shows 200-day price index versus market and versus industry, price trend versus market in the last 30 and last 5 days, volume trend versus the market in the last 30 and last 5 days.

Screen 9—"EARNINGS PER SHARE"—shows the earnings per share in the last 12 months and the last fiscal year, percent of change last year, the fiscal year to date and last 12 months, and the 5-year growth rate.

Screen 10—"REVENUES and RATIOS"—shows last fiscal year revenue and earnings, profit margin, return on equity, debt to equity ratio, and current ratio.

Screen 11—"SHAREHOLDINGS"—shows the market value, last out. shares, insider net trade, short interest, short interest ratio and fiscal year ending date.

Screen 12—"DIVIDENDS"—shows the current dividend rate and yield, the current and 5-year average, payout, and latest dividends.

VI. THE BOOKSHELF

As a member of The Source, you will receive a magazine called SourceWorld that will tell you about features and give tips on how to use them. Source officials also regularly update the official user's manual and you'll be alerted when new editions are available.

In addition, books on computer communications abound. Here are some that we have on our bookshelf.

Smarter Telecommunications, by Charles Bowen and Stewart Schneider, Bantam Books, $12.95. A general guide to the wonders of information services, databases, and computer bulletin boards.

How to Get the Most Out of CompuServe, by Charles Bowen and David Peyton, Bantam Books, $14.95. An on-line tour guide to that consumer information service.

The Computer Phone Book, by Mike Cane, New American Library, $9.95. An annotated listing of on-line services, with special treatment of hundreds of private computer bulletin-board systems.

The Microcomputer User's Guide to Information Online, by Carole Hansen, Hayden Book Co., $14.95. Examples and case studies from some of the top services, with special treatment of commercial databases and vendors.

The Complete Handbook of Personal Computer Communications, by Alfred Glossbrenner, St. Martin's Press, $14.95. One of the first commercial books devoted to dial-up services.

Omni Online Database Directory, by Mike Edelhart and Owen Davies,

Collier Books, $10.95. Directory and evaluation of more than one thousand databases.

The Computer Data and Database Source Book, by Matthew Lesko, Avon Books, $14.95. Encyclopedia of commercial and public databases.

Confessions of an Infomaniac, by Elizabeth M. Ferrarini, Sybex Computer Books, $16.95. A whimsical look at romance on-line.

Using Computer Information Services, by Jeff Williams and Larry Sturtz, Howard W. Sams. An overview of consumer information services.

How to Get Free Software, by Alfred Glossbrenner, St. Martin's Press.

The Joy of Computer Communication, by William J. Cook, Dell Books, $5.95.

Databasics: Your Guide to Online Business Information, by Doran Howitt and Marvin Weinberger, Garland Publishing Co.

VII. REACHING US ON-LINE

Now that you're *really* on your own, we want you to know that you can stay in touch with us. We're only a SourceMail away. If you would like to write to Charlie, send SMail to BBZ824. Dave's Source address is BBP112.

Index

ABOUT THE AUTHORS

CHARLES BOWEN AND DAVID PEYTON have been writing about the electronic community since 1982. They wrote *How to Get the Most Out of CompuServe* (Bantam Books, 1984) and have written articles and columns for an assortment of computer magazines and journals.

Bowen was co-author with Stewart Schneider of *Smarter Telecommunications* (Bantam Books, 1985). He is a freelance writer and a former newspaper writer and editor. He and his wife, Pamela, live in Huntington, West Virginia.

Peyton is a feature writer and columnist for the Huntington Publishing Co., where he has worked for nineteen years. He also is a columnist for Gannett News Service. He lives in Huntington with his wife, Susan, and their fourteen-year-old son, Davy.